Praise for CLOSE ENCOUNTERS WITH HUMANKIND

"Sang-Hee Lee has posed some of the big questions in human evolution and has written brief, clear, and jargon-free essays about them. Her informal, clear style and commonsense explanations are delightful and engaging. Brava!" —Pat Shipman, professor of anthropology and author of *The Invaders*

"In this insightful collection, Sang-Hee Lee shows herself to be a gifted storyteller, breathing new life into the old bones, with both the intimate knowledge of a practitioner and a dose of cross-cultural scientific sensitivity. A pleasure to read!"

—Jonathan Marks, professor of anthropology and author of *Tales of the Ex-Apes*

"I have struggled for years to find reading materials that cover the basics of human evolution without overwhelming students with dry, technical minutiae. At last, a book that does this! This book will appeal not only to students, but to anyone who is curious about the world."

—Dean Falk, coauthor of *Geeks, Genes, and the Evolution of Asperger Syndrome*

"A terrific introduction to the story of human evolution for someone who is just beginning to explore the field, but even those familiar with the material will find a lot of new information here. An appealing, factual, and entertaining book on the ever-fascinating topic of human evolutionary history."

—Wenda Trevathan, author of *Ancient Bodies, Modern Lives*

"Lee uses quick-take chapters and a congenial tone to give us an overview of what paleoanthropology has discovered and how it works. . . . Think of it as visiting some really old friends." —*Library Journal*

CLOSE ENCOUNTERS WITH HUMANKIND

A Paleoanthropologist Investigates Our Evolving Species

SANG-HEE LEE

with Shin-Young Yoon

W. W. NORTON & COMPANY

Independent Publishers Since 1923

NEW YORK • LONDON

Originally published in Korea by ScienceBooks Publishing Co., Ltd, Seoul,
a division of Minumsa Publishing Group.

English translation edition is published by arrangement
with Sang-Hee Lee and Shin-Young Yoon c/o ScienceBooks Publishing Co., Ltd, Seoul,
a division of Minumsa Publishing Group.

Printed in the United States of America
First published as a Norton paperback 2019

For information about permission to reproduce selections from this book,
write to Permissions, W. W. Norton & Company, Inc.,
500 Fifth Avenue, New York, NY 10110

For information about special discounts for bulk purchases, please contact
W. W. Norton Special Sales at specialsales@wwnorton.com or 800-233-4830

Manufacturing by LSC Communications, Harrisonburg
Book design by Fearn Cutler de Vicq
Production manager: Lauren Abbate

Library of Congress Cataloging-in-Publication Data

Names: Lee, Sang-Hee, author. | Yoon, Shin-Young, author.
Title: Close encounters with humankind : a paleoanthropologist investigates our evolving species / Sang-Hee Lee With Shin-Young Yoon.
Description: English translation edition. | New York : W. W. Norton & Company, 2018. | "Originally published in Korea by ScienceBooks Publishing Co., Ltd, Seoul, a division of Minumsa Publishing Group." | Includes bibliographical references and index.
Identifiers: LCCN 2017056766 | ISBN 9780393634822 (cloth)
Subjects: LCSH: Human evolution. | Evolution (Biology) | Paleoanthropology. | Fossil hominids.
Classification: LCC GN282 .L43 2018 | DDC 599.93/8—dc23
LC record available at https://lccn.loc.gov/2017056766

ISBN 978-0-393-35676-2 pbk.

W. W. Norton & Company, Inc., 500 Fifth Avenue, New York, N.Y. 10110
www.wwnorton.com

W. W. Norton & Company Ltd., 15 Carlisle Street, London W1D 3BS

1 2 3 4 5 6 7 8 9 0

Contents

INTRODUCTION: Let's Take a Journey Together 9

1. Are We Cannibals? 19
2. The Birth of Fatherhood 33
3. Who Were the First Hominin Ancestors? 47
4. Big-Brained Babies Give Moms Big Grief 59
5. Meat Lovers R Us 67
6. Got Milk? 77
7. A Gene for Snow White 85
8. Granny Is an Artist 93
9. Did Farming Bring Prosperity? 105
10. Peking Man and the Yakuza 113
11. Asia Challenges Africa's Stronghold on the Birthplace of Humanity 121
12. Cooperation Connects You and Me 131
13. King Kong 141
14. Breaking Back 151
15. In Search of the Most Humanlike Face 161

16. Our Changing Brains 169

17. You Are a Neanderthal! 179

18. The Molecular Clock Does Not Keep Time 189

19. Denisovans: The Asian Neanderthals? 199

20. Hobbits 207

21. Seven Billion Humans, One Single Race? 217

22. Are Humans Still Evolving? 227

EPILOGUE 1: Precious Humanity 235

EPILOGUE 2. An Invitation to an Unfamiliar World of Paleoanthropology 241

APPENDIX 1: Common Questions and Answers about Evolution 247

APPENDIX 2: Overview of Hominin Evolution 255

Further Reading 265

Index 287

CLOSE ENCOUNTERS WITH HUMANKIND

INTRODUCTION

Let's Take a Journey Together

In 2001, I was about to start a new chapter in my life as an assistant professor in the Department of Anthropology at the University of California, Riverside. I was planning to ship everything I had to California, including my car, and then to travel elegantly on an airplane (all possible because UC Riverside was covering my moving costs).

That plan was soon thwarted. My adviser from graduate school strongly recommended that I drive across the country instead. Of course, I objected. Strenuously. I wanted to get to California as soon as possible, to get settled as soon as possible. But that was only a superficial reason; to be frank, the endeavor terrified me. After a long discussion, I gave in to my adviser's convincing argument that this would be the only chance in my life to experience and feel intimately what America was about.

I was reminded of a book that had left a strong impression on me, *Travels with Charley* (1962) by John Steinbeck. I read it after I graduated from college, while preparing to start graduate school in the United States. Steinbeck took his dog Charley and drove across the country, agonizing over what it meant to be an American and what the nation was made of. He wrote frankly about the problems poisoning America, including racial inequality. From his depiction, it was not surprising at all that the civil rights movement broke out in the 1960s, shaking the whole country to its core. And this book made a deep impression on me as I was starting my life in the United States in 1990.

When I left Korea, there was little interest in multiculturalism and diversity among Koreans. To someone like me who had a vague and simple understanding about two races, black and white, the multidimensional aspects of race and the deep-rooted tension between races were quite foreign. So, when I took my adviser's suggestion to drive across the country, I decided to make the most of the experience by talking with people everywhere I went, seeing and feeling everything. I packed a voice recorder, along with the rest of my belongings that were not shipped, into my 1994 Dodge Voyager minivan (a reliable car, but one with manual windows, no air-conditioning, and not even a tape deck) and set out.

I decided on a few guiding principles before I started the cross-country drive. First, I would stay away from freeways and use local roads as much as possible. I didn't have a portable phone back then, so I bought an emergency phone that could only call 911, and a charger that could charge on a car battery. I bought a box of bottled water, a box of crackers, and some simple clothing and toiletries. I felt a little like Captain Janeway from *Star Trek: Voyager*, one of my favorite TV shows. "To boldly go where no one has gone before," I thought, as I left my home and hit the road.

I started from the confusingly named Indiana, Pennsylvania—a suburb of Pittsburgh—where I had spent a year as a visiting assistant professor at Indiana University of Pennsylvania. Indiana, PA, is best known as the birthplace of the actor Jimmy Stewart. During my time there, it was a midsize town that had been declining with Pittsburgh's steel-mining industry. As men lost their jobs in the steel mines, women became the breadwinners, often with jobs in the service industry. Many people in Indiana were not too happy with the changes in the economic landscape—both in the town and at home—and the university, rising to be the primary employer as the steel-mining industry sank, bore some of that ill will. I spent my entire year there just waiting to leave, hoping to teach at a larger school in a more welcoming atmosphere. With this kind anticipation for the future, I started my journey across America.

First I stopped in Michigan to say goodbye to my adviser, Milford Wolpoff. He is one of the original proponents of the "multiregional evolution model" for the origins of modern humans. More important, he was my strongest champion and critic, while treating me like family, since I had come so far away from home. I had fallen in love with paleoanthropology—a transdisciplinary field of science and humanities—despite having little science background in high school or college. Without the unwavering support and encouragement from a mentor like Milford Wolpoff, there's no doubt I would have had insurmountable challenges entering the field.

I left Michigan and went to Kentucky to see one of my friends from graduate school. She had come to the United States with her family from Saigon at the end of the Vietnam War, and we had become close, partly because we were both Asians. We had lost touch after she left graduate school to move on, via a completely different life trajectory, to a position as a business executive in a big company and a happy mother of two children.

Many of us go to graduate school to pursue a life in academia, and many of us leave before completing a degree to pursue another path, as my friend did. In the context of my trip across the country, our reunion made me think about roads not taken. What if she had persisted back then and graduated? What if I had left academia sooner and sought an alternative? On the one hand, it takes a lot of courage to change course after you start the path toward a PhD degree. You have to bear the fear of people thinking of you as a loser and a quitter. But staying on is no light feat either. In the end, no road is easy. Or, rather, all roads are beautiful.

I started out on my cross-country drive with a big ambition. But as I passed through Kentucky, Illinois, Missouri—driving through an endless flatland—I started to get exhausted. After a brief, cool moment early in the morning, the end-of-August sun soon would become stiflingly hot. Without any air-conditioning, I had to drive with an open window. Perpetually heading west, I was almost always facing the sun, and my left arm was growing darker every day. Sunscreen was useless. The local roads were quiet, with just one or two cars passing me on the other side the whole day. The only entertainment source, the radio, was playing country and western music, station after station. With the hot air coming through the open window, riding in a hot car and listening to slow country music, I felt almost like my brain was melting. There's a reason why people on long trips play exciting, simple music.

After driving all day and once the sun started to set, I would stop at the nearest motel.

"Do you have vacancy?" I would ask.

Then, usually a middle-aged tall woman at the front reception desk would look at me suspiciously and ask, "You sure you're by yourself?" The way she shot furtive glances past me, I suspected that she thought I was renting a room "for one" and then sneaking a bunch of people in.

I would grab an easy dinner, come back to my room, watch some TV, wash myself, then go to sleep. Morning would come, I would eat the continental breakfast available at the motel, pay the bill, and get back on the road.

I barely spoke a word all day. Perhaps a couple of words during check-in and check-out at motels. Everywhere I went, I was acutely aware of how different I looked. Everyone else was whiter and bigger than me. I shrank every day, little by little. I didn't want to speak to anyone, partly because I was suspicious of strangers (who didn't really invite me to strike up a conversation anyway), and partly because I was just getting very tired of the whole project. Once every two or three days, I would buy postcards from a gas station or convenience store and send updates to my parents and friends in Korea.

After I passed through Kansas, known to be flatter than a pancake, a huge vista appeared in front of me: the Rocky Mountains. The Rockies are a tumultuous landscape. The road continuously curved, climbed, and dropped through peaks and valleys, and I had to be on my guard. I thought of the many pioneers who had tried to pass this way by wagon but had failed (including, of course, the famous Donner Party who got stranded in the Sierra Nevada and had been forced to resort to cannibalism to survive).

Finally, I crossed from Nevada into California. The first place I visited in California was Calico, formerly a bustling silver-mining town and now nothing more than a washed-up tourist attraction. Calico saw its heyday during the silver rush of the 1880s, with approximately five hundred silver mines operating for twelve years. After the silver price crashed in the mid-1890s, Calico was deserted and turned into a ghost town.

Actually, Calico occupies a noteworthy place in the history of paleoanthropology. In 1960, Louis Leakey, famous for his spectacular hominin fossil discoveries in Africa, pinpointed Calico as the earliest settlement of the first Native Americans, and started an excavation project. Leakey probably wanted to add an American chapter to his great success story from Africa. Instead, the excavation, which started with widespread media attention and public speculation, ended without any noteworthy discoveries, and Leakey left the site. It's still a matter of debate whether the "stone tools" discovered in Calico were human-made tools or just a product of natural rock breaks.

I started in Pennsylvania, drove 3,500 miles through ten states in sixteen days, and arrived in California a few days before September 11, 2001. Just like that, America was no longer a place where a strange-looking foreigner could slowly drive around the countryside in an old minivan.

With my journey over, it was time to start my new life as a professor. I worked hard to prove myself as more than just a diversity hire, though I suspected that may have been the reason I'd been offered the position.

Being a professor was challenging. Having grown up in a culture where the king, father, and teacher are considered the holy trinity, I found it difficult to get used to an environment where students might consider professors to be their friends and think nothing of expressing differing opinions. Of course, I had experienced a little of this first-hand as a graduate student at the University of Michigan, but it was totally different as a professor.

Initially, I held classes the way I was taught in college. I erroneously thought that students would be grateful for the knowledge I imparted to them, and that they would absorb my words like sponges. The students, however, had different ideas. They sat in the classroom, arms crossed, challenging me to earn their respect and attention. It was a completely different atmosphere from my college experience in Korea, where being a part of a university, as a student or faculty member, was a mark of pride and demanded respect from others.

I realized much too late that I did not have a natural talent for teaching. In despair, I put all my efforts into research. The years passed, I settled in as best I could, and eventually I received tenure. Then, out of the blue one day, I was contacted by Shin-Young Yoon, a science journalist in Korea, who proposed I start a column about human evolution in *Gwa Hak Dong A*, a Korean science magazine for the general public. Intrigued, I accepted, and began writing essays for the general public on various interesting topics in human evolution. As I continued the series, I realized how limited and ineffective it is to communicate in a simply unilateral and authoritarian manner. Yet this is how I had been teaching.

I started to tell stories to my students the same way I told stories in my essays for the Korean magazine. Then something miraculous happened. I began to feel passion and enthusiasm about a large course I was teaching regularly: Introduction to Biological Anthropology.

Some professors prefer an intimate class, exploring a narrow topic in great depth with a small number of interested students. I used to be one of them. But now I prefer large classes. Among the hundreds of students sitting in large introductory classes, some are taking the course only because they have to fulfill a general education requirement. Before, I used to get discouraged looking at the bored faces of the students filling the back half of a large lecture hall. Now I love the challenge of sparking some curiosity in those same students, of trying different ways to reach them, of motivating them to want to learn more. It keeps the course fresh for me, even though the content stays relatively the same. And inevitably, some of those students who came to take the course because they had to, not because they wanted to, decide to change their majors to anthropology.

I've found that the same topics that interest young students also interest adults. Of course, this is not surprising: where we came from, how we lived, or why we look the way we do are some of the fundamental questions we all wonder about at some point in our lives. In newspapers and magazines, new discoveries of human ancestral fossils always get attention. The one lesson that is repeatedly conveyed in all those stories of human evolution is that there is no right answer and there are no bad questions: human evolution is an ever-changing field.

What is accepted as the right answer at one time can be challenged by new data and a new hypothesis at another time. A trait that is evolutionarily advantageous, that has an adaptive advantage, is ultimately a product of chance. Individuals with a trait that happened to appear, that happened to receive a benefit in their equally random environment, got to leave behind a few more offspring than others who did not have that particular trait. But a trait that is advantageous at one point in time is not advantageous forever. Everything changes.

Such is the case through the long evolutionary history of humanity. Sure, there have been major advances in our development, steps that have moved the human race forward. Upright walking is one, a bigger brain is another, and dependence on culture is yet one more. But when we look more closely at the human journey, we see not a straight line, but a curvy, winding river. Humanity did not agonize over the best long-term course for development. We proceeded by making the best decision possible at that moment, within our specific environment.

In high school, I was assigned to be in the humanities track instead of the sciences track, because the aptitude test said so. Yet I ended up studying a field that was both humanities and sciences. I used to think I had no talent for teaching, but now I think I'm a great fit for the profession. Not for a moment, however, do I think it will stay this way in the future. The environment I'm in might change, and so will my body and heart. It was the same way for those sixteen days I spent driving across the country. The only thing I could do was decide which road to take while drinking my coffee in the morning, and then drive west for the rest of the day.

The twenty-two stories in this book are inspired by interactions with students in class, and moments that I experienced directly and indirectly. The stories are written in an essay format that is meant to be fun and intriguing. Many of these chapters start with a question I was asked, and some begin with a story that inspired me to delve further into a topic of interest. I have tried to make these topics completely understandable to someone without a background in paleoanthropology. Feel free to start from the beginning and read through all the chapters in sequence, or start anywhere you like and read in random order. Feel free to close the book after you're done, or investigate further by exploring the references provided in the Further Reading section at the end of the book. I simply invite you to join me in this fun and exciting journey tracking the origins of humanity.

Sang-Hee Lee

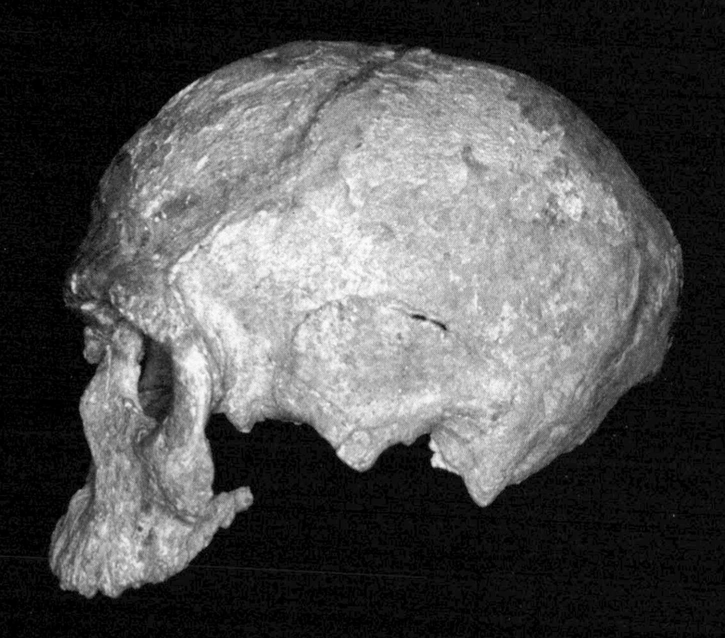

Fossil skull of early modern human, *Homo sapiens*, discovered in Jebel Irhoud, Morocco.
(© Milford Wolpoff)

CHAPTER 1

Are We Cannibals?

In the movie *The Silence of the Lambs* (1991), starring Anthony Hopkins and Jodie Foster, Hopkins plays Hannibal, the cannibal. This is one of the few movies I bought a ticket for and then, in the middle of it, left the theater. Going in, I had a vague idea of what the movie was about. I felt a little nauseated by the movie's premise, but I figured I could handle it. I clearly overestimated my gumption, and ran out after one too many gruesome scenes.

Thus it's quite ironic that some years later, I would come to be known as the "cannibal expert," albeit briefly. It was spring of 2007. Someone called me at my office.

"Hi, I am so-and-so [I can't recall his name] in Hollywood, working for *E! News*. I have a couple of questions for you, since you are the expert in cannibalism. If someone snorts someone else's ashes, would you call that cannibalism?"

"Huh?"

"Yesterday, Keith Richards of the Rolling Stones said that he snorted his father's ashes. You know who the Rolling Stones are, right? So I wanted an expert opinion on this and Googled 'cannibalism.' Your name came up as the first hit. I didn't know that we had a cannibal expert so close! I'm so happy to know you." I was mortified to find out that my name was coming up as a top Google search result for "cannibalism."

I had taught a class on cannibalism a couple of times, mostly because some students were interested in the topic. Then the class was featured in the *Chronicle of Higher Education*. Colleagues at the university started to tease me as the "cannibal professor." Some sent me newspaper clippings about cannibalism. At that time there was also an ongoing cannibalism court case in Germany, which I had used as an example in my course. Apparently, someone in Germany had advertised to recruit people to be eaten, then proceeded to kill and cannibalize those who answered the ad, after signing a contract with the victims.

Given my general aversion to the idea of cannibalism and only this tenuous connection to the subject, I was understandably agitated to get this phone call from a Hollywood journalist. When I answered the phone, I first wondered whether it was a prank call, but I went ahead and gave him my expert opinion anyway. Cannibalism depends on how it's defined. In some cultures, imbibing the ashes of ancestors is a custom to show respect for the deceased; the Yanomami, for example, are known for this behavior. Many anthropologists would classify this practice as a form of cannibalism. There was no way to know whether Keith Richards was moved to snort his father's ashes by a comparable respect toward his late father, or if his act of snorting was categorically the same as consumption. My interview with the journalist was published in print and on the web; some of my friends called me, awestruck that my name appeared in the same paragraph as the legendary guitarist.

The alleged story about Keith Richards does raise a very interesting question about human nature itself: Are we cannibals? Humans are extreme omnivores; to say there is nothing on Earth that humans do not eat would be only a slight exaggeration. It might not be a stretch to think that there is a tribe somewhere on Earth that serves up fellow humans at their dining table on a regular basis. Movies often depict people who become lost in the jungle, are caught by cannibals, and then dramatically escape right before being boiled or roasted to be eaten. When asked "Who are cannibals?" most people refer to some idea of "primitive" peoples living in the jungle. We often think that we civilized people could never be cannibals, but perhaps in a faraway land there are savages, inferior to us, engaging in such shocking dietary customs as cannibalism.

I will return to the question of whether there really are cannibals. First let's talk about another group that some anthropologists suspect were cannibals. This group is quite significant in discussions of cannibalism in modern-human populations. Interestingly, the group did not live in a faraway place, nor does it occupy a point far away in time. It is the now deceased lineage of modern humans (*Homo sapiens*) known as the Neanderthals.

Our cannibal relatives?

Krapina, Croatia, is a cave site that was excavated in the early twentieth century. The Krapina cave is famous because dozens of Neanderthals were found buried there. Among the remains were many young women and children, all of whom shared some intriguing characteristics. First, none of the individuals were complete specimens; there were only fragments of each. In particular, there were fewer facial and cranial bones than would be expected. In addition, the bones had peculiar cut marks. What did this mean?

Paleoanthropologists interpreted all this as evidence of cannibalism. In the early twentieth century, we imagined Neanderthals as brutal savages, violent and uncivilized. Some of you reading this right now may still have this brutish impression of our ancestors: hairy, stooped and short, with a violent tendency, somewhat akin to apes living in the African jungle. This negative impression created a bias toward "evidence" of Neanderthals as cannibals, which became a widespread opinion in the first half of the twentieth century.

Things began to change, however, in the second half of the twentieth century. Some anthropologists started to voice the opinion that there had never been cannibals, Neanderthal or otherwise. Then, an interesting study was published in the 1980s by Mary Russell. Russell, then an anthropologist at Case Western Reserve University, had come up with a new and clever way to test whether Neanderthals had, in fact, been cannibals.

Russell knew that many paleoanthropologists assumed that Neanderthals really did kill and eat each other, and were therefore quick to attribute the cut marks on the fossils to the butchering process. But was another explanation possible? Russell hypothesized that there might be an alternative account for the marks: a "secondary burial." In a secondary burial, a deceased person, already initially buried, is exhumed after a certain period to have the bones cleaned, and then buried again. In certain parts of Korea, secondary burials were practiced until recently. The cleaning of bones for burial was also an ancient practice observed in some Polynesian and Native American cultures. In these cases of ritual reinterment, the cut marks are not from butchering, but from the detailed cleaning and reburial of bones.

To test whether reburial could explain the evidence found at the Krapina Neanderthal site, Russell collected cut-mark data from archaeological sites with confirmed butchered remains and from other sites confirmed as secondary burial sites. First, she collected bones with cut marks from big-game hunting and butchering from the Upper Paleolithic. Then, she examined the bones at a Native American ossuary with distinct cut marks from secondary burials. She compared the cut marks from these two different sites with those from the Krapina Neanderthal site.

As you might guess, the cut marks from the Krapina cave turned out to be quite different from the butchering marks made on animal remains, and more similar to those from the secondary burial sites. In particular, the Krapina cut marks were on the ends of the bones. This pattern was very similar to marks made in the secondary burials of Native Americans, and clearly these marks were not the kind that would result from butchering meat for consumption.

It is easy to understand this contrast if we think about the process of secondary burial itself. Usually at the time of a secondary burial, the body is decomposed substantially, and bones can be cleaned just using a knife. Generally, most cleaning needs to be done at the ends of the long bones (where the joints are), leading to a concentration of cut marks at those points. In contrast, cuts made from butchering leave marks in the middle of bones because the meat (muscle) has to be cut off the bone for consumption and the midpoint is where the muscle is attached. Russell's study showed that the Neanderthal cut marks most likely aligned with funerary practices, not with butchering. Therefore, the Krapina cave cut marks could not be used as evidence of cannibalism in Neanderthals.

Are "cannibals" all just misunderstood?

In the 1980s, when Russell published her research on the Krapina Neanderthals, the idea that there are no cannibals (and probably never were) was slowly spreading among anthropologists. Some argued that the idea of cannibalism was just a misunderstanding or a bias. In fact, the very use of the word "cannibal" is the result of a misunderstanding by Christopher Columbus. Upon arriving in the West Indies in the fifteenth century, Columbus mistook where he landed to be India, and he mistook the people he met to be Mongolians, also commonly referred to as the "descendants of Khan." Thus he called them "cannibas."* He then sent a report back to Europe saying that "the Cannibas ate people."

Columbus's story spread quickly through all Europe, and "the Cannibas" became generalized to "cannibals." Europeans were fascinated to learn that cannibals, who until then had existed only in legends and myths, were real. European countries began to compete against each other in securing colonies, and they sent missionaries, explorers, and anthropologists to collect stories about cannibals from faraway places and publish them in articles or books as a form of popular entertainment. Cannibalism became one of the required characteristics of "primitive peoples."†

By the latter half of the twentieth century, however, a very different story had emerged. Close scrutiny of the books and reports about cannibals revealed that many of the cannibal stories had no sound basis. Numerous "reports" turned out to be just rumors. William Arens, an anthropologist at the State University of New York at Stony Brook, closely examined all the cannibal records and posited an explanation of the origin of cannibal rumors in his book *The Man-Eating Myth* (1979). The source of the cannibal rumors was often recorded testimony from a member of a neighboring or a competing indigenous group, who would tell curious European travelers where one might encounter these "cannibals." The testimony often sounded something like this: "We don't do such things, but the people who live on the

* Another argument is that "cannibal" comes from *caríbal*, the Spanish word for "elected chief."

† Cannibalism marked people for slavery. The Spanish issued a doctrine saying that only people known for cannibalism would be enslaved.

other side of the forest are ruthless cannibals. I was almost caught and eaten myself, and courageously escaped." Certainly, no European who wrote about cannibals during the early colonial periods was a direct eyewitness to cannibalistic behavior.

Arens's close examination of these statements led the field to suspect that there was no anthropological basis for claims of historical cannibalism, at least as a part of a regular, normal diet. But can we conclude that there has never been any cannibalism whatsoever in human history? Not necessarily. In fact, certain populations of people, albeit few and far between, are known to have engaged in cannibalistic behavior. Most notorious are the Fore of Papua New Guinea. The Fore people were unknown to the outside world until Australian officials arrived on the island of Papua New Guinea in the 1940s to conduct a demographic census of its inhabitants. By the 1950s, Australians had built a security station and vehicle roadways into the region. Shortly after, anthropologists and missionaries started to arrive as well.

The outsiders immediately took note of cultural customs of the Fore people, including their consumption of the dead. Considered to be cannibals, the Fore were pressured to stop the behavior. The cannibalism of the Fore, however, was highly ritual and tied to their unusual funerary practices, which involved the partial consumption of deceased kin. Among the Fore, when a relative passed away, the body of the deceased was cleaned by matrilineal kin through a unique process that is still unknown in any other human population. It might be a little gory, but let me describe it.

First, the hands and feet of the deceased were cut off; then the arms and legs were defleshed for meat. Then the brain was taken out, and the stomach was cut open and the intestines removed. Following these steps, male kin were given the muscle meat to consume, while women consumed the brain and the intestines. In addition, during this process of cleaning the deceased body, onlooking children were encouraged to partake in this ritual honoring the dead.

Although this ritual was widespread in the past, the Fore no longer practice it. But the question remains, why on earth were they doing something considered so abhorrent to so many other people? The answer lies in their particular funerary belief system. By internalizing the dead, the Fore believed their deceased kin would again become part of the living and continue to exist in the village. This may sound outlandish to some, but the belief itself is not so unusual. In fact, many other cultures and religions have a similar version of this practice. For example, the Yanomami of the Amazon mix the cremated ashes of their deceased into a gruel, and all the village people (who are also relatives) consume it. Moreover, the Christian Eucharist, or communion, is based on the belief that one is (metaphorically) devouring the flesh and the blood of the savior Jesus Christ. These cultural practices all convey the same message: "Do this in remembrance of." Therefore, it is not unusual to find that behind the gory cannibalism of the Fore lies a love for kin that we all share.

Of course, not all cannibalistic behavior is necessarily affectionate. Some cannibalism originates from conflict. The drinking of blood or eating of a captured enemy's heart during warfare is one such example of malicious cannibalism. Here, the purpose is to eradicate enemies by consuming them. Notably, such acts are only described in historical records; there are no eyewitness accounts of this practice in modern history either.

Whether the practice is motivated by love or by hatred, one thing is clear about cannibalism among humans: no human population eats other humans as part of a regular diet. In other words, eating another human is never part of the repertoire of normal behavior. The cases I've discussed are all examples of ritual symbolistic behavior or an occasional cultural custom, but not really cannibalism. Cannibalistic conduct originates not from hunger, but from love or hatred—both thoroughly human emotions—expressed through ritualized behavior.

There may be cannibalistic behavior, but there are no cannibals

Let's return to paleoanthropology. Applying the clever method of Mary Russell, archaeologists and paleoanthropologists decided to revisit several other discovered remains, looking for cannibalistic behavior in the past by using the same cross-comparison of cut marks on bones. In 1999, cannibalistic cut marks (cuts toward the middle of bones rather than the end) were found among Neanderthal remains discovered in Moula-Guercy, France. Seemingly cannibalistic cut marks were also found in Atapuerca, a Middle Pleistocene (approximately 780,000–120,000 years ago) site in Spain that predated the advent of Neanderthal occupation. And similar cut marks were found, too, on human bones uncovered at ancient Native American sites in the United States.

The American discovery started a heated debate about the implications of historical cannibalistic behavior. Whether the indigenous ancestors of Native Americans were cannibals was a sensitive topic, involving unresolved emotional and political tensions between contemporary Native Americans and the perceived descendants of the Europeans who took indigenous lands by way of conquest and genocide. Native American communities took the charge of ancestral cannibalism quite personally. The debate shifted to the political implications of calling ancestral Native Americans cannibals, rather than focusing on the empirical data.

Then, critical evidence was presented in 2001 that seemed to settle the debate. A protein found only in human skin was discovered in fossilized human feces (coprolite) at Anasazi, a Paleoindian site in Colorado. The fossilized evidence was considered a smoking gun, indicating that at least some form of cannibalism had taken place in this location at some point in time.

We should keep in mind, however, that evidence of cannibalistic behavior does not prove the existence of cannibals. As in the example of the Fore and others discussed earlier, it is clear that ritualized cannibalistic behavior has existed throughout human history. Cut marks found at even earlier sites in France and Spain, and in other southern Paleoindian sites, might also be evidence of such behavior.

Moreover, in modern times there are occasional instances of acceptable cannibalistic behavior under extreme circumstances. A famous example is the case of the Uruguayan rugby team, who became lost after their plane crashed in the Andes in 1972, and survived by eating the dead bodies of their colleagues. This incident became the basis of the movie *Survive!* (1976). And, of course, members of the Donner Party survived by eating their deceased companions after being trapped in the Sierra Nevada for four months. Can we call these people, who had to survive through extraordinary and exceptional circumstances, cannibals? If the Chilean miners trapped in a collapsed mine in 2010 had been forced to resort to such extreme measures to survive, I doubt we would be measuring them with an ethical ruler and calling them cannibals.

In the same way, the hominin fossils from the past demand of us a more creative and imaginative interpretation. Could they have eaten their fellow hominins to remember them? Or perhaps to enact revenge during warfare? Or as the last extreme measure to survive under the extraordinarily harsh circumstances that befell them during the Pleistocene (also known as the Ice Age)?

We cannot make assumptions about our past beyond the conclusions that the archaeological and paleoanthropological data support. Indeed, there is evidence of ancient cannibalistic behavior—but we cannot call those who engaged in that behavior cannibals for sure.

EXTRA
KURU, THE STRANGE DISEASE OF THE FORE

One reason the Fore's cannibalism became widely known was the advent of a strange disease in the 1950s. Upon learning about the spread of an unknown disease among the Fore, Australia sent an investigative team of doctors, who reported that afflicted women complained of feeling extremely weak and could not stand; instead, they could only lie down in bed, eating very little. By the end of the progression of the disease, afflicted individuals would experience tremors and convulsions throughout their bodies and ultimately die. Because of the tremors, the disease was called *kuru*, which meant "shaking" in the indigenous language. It was also called the "laughing disease" by some, because patients would fall into fits of nonstop laughter.

Kuru has a very long incubation period, normally between five and twenty years, but sometimes as long as forty years. For example, the last reported person to have the disease died in 2005, but he was infected in the 1960s. After the incubation period, a patient usually dies within a year of the symptoms' first appearance. During the first few months of the disease's progression, the muscles gradually loosen and cannot be controlled, leading to slow immobilization. Language function and bowel movements cease, and swallowing food or drink becomes impossible. Finally, death usually arrives through secondary complications of pneumonia, starvation, or infection from bedsores.

The horrible disease was a strange mystery to the Australian doctors. There seemed to be no identifiable reason for its occurrence. Enter Daniel Gajdusek, a member of the Australian medical team. He came to know of kuru while conducting research on diseases that seem to affect predominantly indigenous populations living in remote areas. While reviewing the reports about kuru, Gajdusek read that the Fore were known for their cannibalistic practices. He wondered whether there was a relationship between kuru and cannibalism. In particular, he noted that women and children—who made up the majority of kuru cases—ate the *brains* of the dead.

Gajdusek suspected that the source of the kuru disease resided in the soft tissue of the brain. In an experiment, he transplanted some brain tissue of a patient who had died of kuru into a chimpanzee. Within two years, the chimpanzee showed the same symptoms of the kuru disease. Further experimentation by Gajdusek revealed that the pathogen for kuru is indeed located in brain proteins called prions, and one could contract the disease by eating infected meat. Prions are misfolded proteins that can induce other normal proteins to misfold. The scientific community had hypothesized that proteins could possibly be such a unit of inheritance, but the theory was yet to be empirically found. The research that Gajdusek conducted on kuru documented the existence of a prion disease for the first time.

Kuru is part of a class of diseases known to plague other mammals and humans by causing complete neurological degeneration. Unlike cancerous cells, which multiply uncontrollably through regular cell division processes, prions transform the cells around them. Several prion-related diseases, such as mad cow disease and Creutzfeldt-Jakob disease (CJD), have since been discovered, but Gajdusek's research is acknowledged as some of the most groundbreaking in medical history. Gajdusek was recognized for his discovery of prions with a Nobel Prize in Physiology or Medicine in 1976.

Aside from Gajdusek, no one suspected that kuru would spread through cannibalism, because the Fore usually did not eat those who had died of diseases. But kuru was an exceptional case. The Fore thought of kuru as a disease of the mind, not of the body, so they ate those who died from kuru. Between the late 1950s and the early 1960s, more than a hundred people died of kuru (the patient discussed above who died in 2005 was the last patient). Currently, scientists hypothesize that the kuru epidemic started with the funerary rituals for one person who must have had kuru, which was endemic to the population. The disease spreads not only from eating the infected brain, but through open wounds. The women who continued to clean the dead bodies even after being cut during the cleaning process were probably infected this way as well, increasing the contagion rate.

Neanderthal fossil humeri (arm bones) from Krapina, Croatia. (© Milford Wolpoff)

CHAPTER 2

The Birth of Fatherhood

There is no question that the human family is unique: human families involve adult men. This simple fact sets humans apart from almost every other primate. The mother-infant dyad is a near-universal basic unit in primate social groups. Females give birth and take care of their young until they can be on their own. Although other females will help in taking care of the baby, the majority of the responsibility lies with the mother. In contrast, human mothers rely on others to help with childcare by spending time with their children or by providing supplemental resources. And in many cases, that additional contribution comes from fathers. Yet for many other animals, reproductive success for males consists of getting as much access to as many females as possible, with little to no concern for raising offspring. What led to the unique human model of family? Let's explore this question by starting with our primate ancestors.

Gorillas and chimpanzees use different mating strategies

Nature does not like to waste energy. And that goes for mating and having offspring, too. At least in theory, a male's reproductive capacity is infinite. His number of sperm is substantial, he can replenish his supply, and he can impregnate other females right after impregnating one female. Thus, it follows that the optimal goal for a male is to transfer as much sperm as possible to as many females as possible.

In contrast, a female's reproductive capacity is limited. The number of eggs she can release is small and fixed, and once a female becomes pregnant, no additional pregnancies are possible during her periods of pregnancy, childbirth, and breastfeeding. For females, therefore, the optimal goal is to choose well, selecting sperm of the highest possible quality, during their fertile period. One could say that males are for quantity and females are for quality, as far as reproductive endeavors go. Since males and females have a conflict of reproductive interest, they need to employ diametrically opposed strategies. It is rather strange to posit that males and females, who have to coordinate their efforts to reproduce, must also engage in competitive strategies to best pass on their genes.

For most apes, the intensity of male-male competition depends on the energy involved in raising offspring, and the number of males available for each female in estrus (the period during which females signal that they will soon be fertile). Generally speaking, the more a female provides for herself and her offspring, the less involved the paternal male will be after mating. If, instead, the female provides less, the male has to contribute that much more energy; hence the male will spend less effort making babies (by competing with other males) and more effort raising them.

What happens if all females reached estrus at the same time, as they do in gorilla communities? In theory, males would not have to spend as much time and energy on guarding mates or raising offspring; they could just approach all females at once, during the period of time they were in estrus. In this scenario, however, all males would be in competition for access to mates at the same time, and every male would have to outcompete other males for exclusive mating privileges. As a result, a substantial amount of energy would be spent, perhaps even wasted, in male-male competition during female estrus.

To prevent such inefficient use of energy, male gorillas (to take one example) often engage in competition and decide on ranks *before* the estrus period, to ensure that during estrus, only high-ranking males will approach females. In this case, it is quite difficult for a lower-ranking male to openly approach females within the group. Since the ranks are already decided, the high-ranking males can make symbolic displays of aggression without engaging in physical competition. This way, males can conserve their energy for mating. This strategy benefits the higher-ranking males, but it is quite detrimental for lower-ranking males, since it means that these males may never have a chance to reproduce.

The real winners of this strategy may be the females, since they do not have to agonize over selecting the highest-quality mate. Instead, only high-ranking males will approach females after the intense competition period to sort out the hierarchy among themselves. The females may be singing the song of triumph!

In this kind of competition among males, the two traits that help determine outcomes are body size and canine size. For males, the bigger they are, the better they fare in physical (and visual) competitions. Gorillas are perhaps most famous among primates for such male-male competitions; male gorillas are much bigger—in body size, skull size, and canine size—than female gorillas. This difference between males and females, called body size sexual dimorphism, reveals the intensity of male-male competition. In short, the more intense the male-male competition is, the bigger the males are in comparison to the females.

Some apes, however, take a different approach. Take a look at chimpanzees, for example. Female chimpanzees do not have synchronized estrus. In other words, different females become fertile at different times—a challenging situation for chimpanzee males. Even for the strongest male, it is extremely difficult to constantly guard a mate at all times against other males (in contrast, gorilla males have to guard a mate only during the relatively short period of estrus).

Unsynchronized estrus is challenging for chimpanzee females as well. Because of the mate-selective pressures associated with unsynchronized fertility, the strategy of female chimpanzees differs from that of female gorillas. Female chimpanzees copulate with as many males as possible. Needless to say, males also copulate with as many females as possible. In chimpanzee groups, males do not compete intensely with each other for rank, and various males may approach any female in estrus.

Consequently, copulations occur throughout the year without a particular peak period. In such a scenario, what does a male do who wants to outcompete others and make sure his genes are transferred to the next generation? The answer lies in his sperm. Chimpanzee males release copious amounts of sperm to compete with sperm from other males. The competitive advantage in this case, then, is to make as much sperm as possible. For this particular strategy to work, bigger testicles are more useful than bigger body size. As you might expect, chimpanzee males and females have minimal body size sexual dimorphism, meaning they do not differ much in body size or head size (although they may differ more in canine size). But chimpanzee males have the largest testicles in proportion to body size among apes.

Mating strategies in humans

Once mating occurs and pregnancy ensues, female chimpanzees do not know whose sperm contributed to making the baby. The father is essentially unknown. And female gorillas, despite the mate-guarding practices from males, don't really know who the father of the offspring is either. Just because a male has a high rank does not mean he has an ironclad monopoly on fathering the next generation. Paternity testing among gorillas has shown that the lowest-ranked males surprisingly have a fairly good chance of reproducing, especially over middle-ranked gorilla males. Why? Because middle-ranked males become stressed by the competitive attention they receive from higher-ranked males, while low-ranked males fly under the radar. During the time that high- and middle-ranked males are concentrating on competition, low-ranked males may be able to court females.*

Neither of the mating strategies adopted by gorillas and chimpanzees give males a clear sense of paternity. Therefore, it makes sense for males to concentrate their efforts on making babies, and invest no effort in raising them. Indeed, that's what we see: no male apes raise babies, and chimpanzees thus don't have the experience of being raised by a father.

As we all know, however, humans are different. Human males, with smaller bodies than gorillas have and smaller testicles than chimpanzees have, take an approach to reproductive success unlike that of any other ape: they focus on childcare.

Let's consider the first bipedal hominins, from 4–5 million years ago. Females who were pregnant or breastfeeding likely had some difficulty moving about. They most likely limited their range of move-

* Sometimes males whose growth is complete delay their secondary sexual characteristics in order to look like pubescent males, and succeed in reproduction. This strategy is not intentional, but rather a result of being exposed to a challenging environment that induces stress hormones, leading to delayed maturation and sexual development.

ment or foraged more for plant-based food sources. Males, on the other hand, would have had free hands and less weight to carry around. They could travel more widely and hunt for animals as well. Males could use acquired food to their advantage as leverage to access females in estrus.

But what about pregnant or breastfeeding females not in estrus? Since ovulation suppression makes females infertile during breastfeeding, there would appear to be no benefit for males to get on their good side. Why wouldn't a male give his food to other promising females instead? The obvious argument is that a pregnant female might carry his genes; provisioning for that female then would have a direct benefit for that male's own reproduction. Of course, there needs to be an assurance that the baby involved is indeed genetically related to the male in question. Humans have tackled this question by taking mate guarding to the extreme, in the form of monogamous relationships.

Male versus female: keep estrus secret!

Let's take a look at the monogamous situation from the female's point of view. It is certainly a benefit to the female if the male continues to bring meat and other resources. Females are fertile for only a couple of days each month, however. How does the female make sure the male keeps bringing food for the rest of the month? One explanation proposed that the female's strategy was to hide estrus and force the male to provide continuously. Females have come up with a marvelous way (evolutionarily speaking) of hiding their estrus not only from males, but even from themselves. As human females, it was argued, we do not necessarily know the precise time of our estrus, and we have our evolutionary ancestors to thank for this. The result? Humans have sex all the time, regardless of fertility cycles, and men end up returning to the same women.

Some scientists believe that the union of one man and one woman, mediated by the exchange of sex and food, led to a package deal in human evolution of sexual division of labor, the nuclear family, and bipedalism—all of which allowed the males to provide for females and children. This theory of human origins is called the Lovejoy model. Owen Lovejoy, an anthropologist at Kent State University, proposed this model in a 1981 paper with the title "The Origin of Man," published in the renowned journal *Science*. His article caused a passionate social response, which became his claim to fame.

Anthropologists have examined the Lovejoy model closely. If the model is correct, early hominins should show signs of bipedalism. And since male-male competition would be weak, early hominins should show smaller differences between males and females in body size. The low level of male-male competition should also have led to smaller canines in males.

Looking at the earliest known hominin when Lovejoy introduced his model, *Australopithecus afarensis*, we see that only half of the predictions were supported by the available data. *Australopithecus afarensis* had canines smaller than those of gorillas or chimpanzees, but bigger than those of modern humans. Sexual difference in body size was also less than what is observed in gorillas, but more than in modern humans. *Australopithecus afarensis* was bipedal. All this evidence might suggest that early-hominin mating strategies were different from those of either gorillas or modern humans.

In 2009, a special issue of *Science* included several papers on a newly discovered early hominin species, *Ardipithecus ramidus*, who existed nearly a million years earlier than *Australopithecus afarensis*. Included in the issue was a paper by Lovejoy's research team, which argued that *Ardipithecus ramidus* was bipedal and had a small body size difference between males and females. So, does this mean the Lovejoy model is correct?

Was Lovejoy wrong?

Lovejoy's provisioning model provoked an especially strong critique from feminist anthropologists. The model implied that the idealized "nuclear family"—a monogamous marriage pair-bond between a man and a woman, with the man working and earning money while the woman stays at home raising children—was supposedly written into our DNA from the beginning of human evolution. In other words, women have, for millions of years, been getting food in exchange for sex with men.

Research in the past thirty years lends support to the idea that the Lovejoy model is wrong. First of all, humans are not the only species engaged in recreational sex during and outside estrus. Dolphins and bonobos (*Pan paniscus*, the closest relative to humans) engage in continual sexual activity, but they do not have nuclear families. This ideal of a nuclear family is most likely a product of capitalism and the market economy, rather than a biological imperative. Lovejoy's model might have been less about human origins and more about men's fantasy of having infinite sex.

Most surprising, humans do not have hidden estrus, contrary to the basic assumption made in the Lovejoy model. Women act differently, knowingly or unknowingly, during ovulation; and men respond accordingly, knowingly or unknowingly. Anthropologists have discovered that during ovulation, women have a higher-pitched voice and lower appetite, experience swelling of the mammary glands, and might (unknowingly) choose to wear clothes that are considered attractive by both men and women. Men (unknowingly) are more strongly attracted to the smell of ovulating women and secrete more testosterone, the male hormone responsible for heightened male sexual activity, as a result.* If one

* Interestingly, men do not exhibit such responses to other ovulating women if they are in a stable relationship.

day, a certain woman looks particularly beautiful all of a sudden, you (I'm addressing the heterosexual men here) might be responding to an evolutionary hormonal calling.

The dawn of humanity and the birth of fatherhood

In human families, men usually play a prolonged role as fathers, putting care, love, money, and time into raising their children. According to the Lovejoy model, they behave this way because the children are genetically theirs. But something is not right with this genetic-calculus argument. Like other male apes, a human male has no direct means of knowing for sure whether he is the genetic father of the children he's raising. A paternity test can take care of that uncertainty, but this is a very recent invention. Moreover, even with today's technology, few men bother to test their children's paternity, choosing instead to believe that they are the biological fathers.

Considering the substantial amount of resources that children need to grow up, one might expect more men to take measures to be certain of paternity, but this is not the case. Fatherhood, then, appears to be a cultural concept, rather than a biologically determined role. In monogamous relationships, men *believe* that the children produced by their union are their own. They accept that they are the fathers of the children born to the woman in that household.

What's fascinating is that this cultural role leads to biological changes. When men get married or become fathers, their testosterone decreases. Maleness, and with it the drive to copulate with many women, is reduced once a man assumes the role of a monogamous husband or father.

Today the Lovejoy model is challenged at its foundational assumptions. Men and women are males and females, but also cultural entities

beyond biology. The birth of fatherhood proves this. Women and men are ultimately cultural human beings.

EXTRA

COUVADE, THE PHENOMENON OF IMAGINED PREGNANCY BY MEN

There are many variations of what fathers look like, across history and cultures. In traditional patriarchies of the past, fathers were removed from the daily lives of their children, providing resources from a remote distance. They often stayed in separate living quarters. Even when fathers and children lived in the same space, fathers would rarely see their children, because they worked long hours. Children stayed with their mothers, but the important decisions were made by fathers. "Stern father, kind mother" was the ideal established in the patriarchal social context.

In the twenty-first century, fathers have changed to become more family oriented and accessible. They accompany pregnant partners during doctor visits, coach women through labor in the family delivery room, and take on a sizable share (optimally, half) of the childcare after birth. Mothers, of course, take on breastfeeding, but fathers can participate by bottle-feeding their babies, changing diapers, or tackling numerous other child-rearing tasks.

Fathers also participate beyond childcare. Surprisingly, a large number of men experience something called a "sympathetic pregnancy"—including morning sickness, weight gain, and stomach pain similar to the feeling of a baby moving inside the womb—when their partners become pregnant. Occasionally, they even feel labor pains next to their partners giving birth. The imaginary experience of pregnancy and labor is not just psychological; it is also clearly physiological. Men whose partners are pregnant mirror the woman's hormonal changes from early pregnancy until postpartum. And sometimes, cultures even induce this sympathetic sharing of pain. In traditional Korean society, to take one example, a woman in labor would get through the pain by holding on to her husband's man bun.

In anthropology, this imaginary or sympathetic pregnancy and labor experience is called "couvade syndrome," which just goes to show that biology and culture have learned to work together to help prepare the father for his new nurturing role.

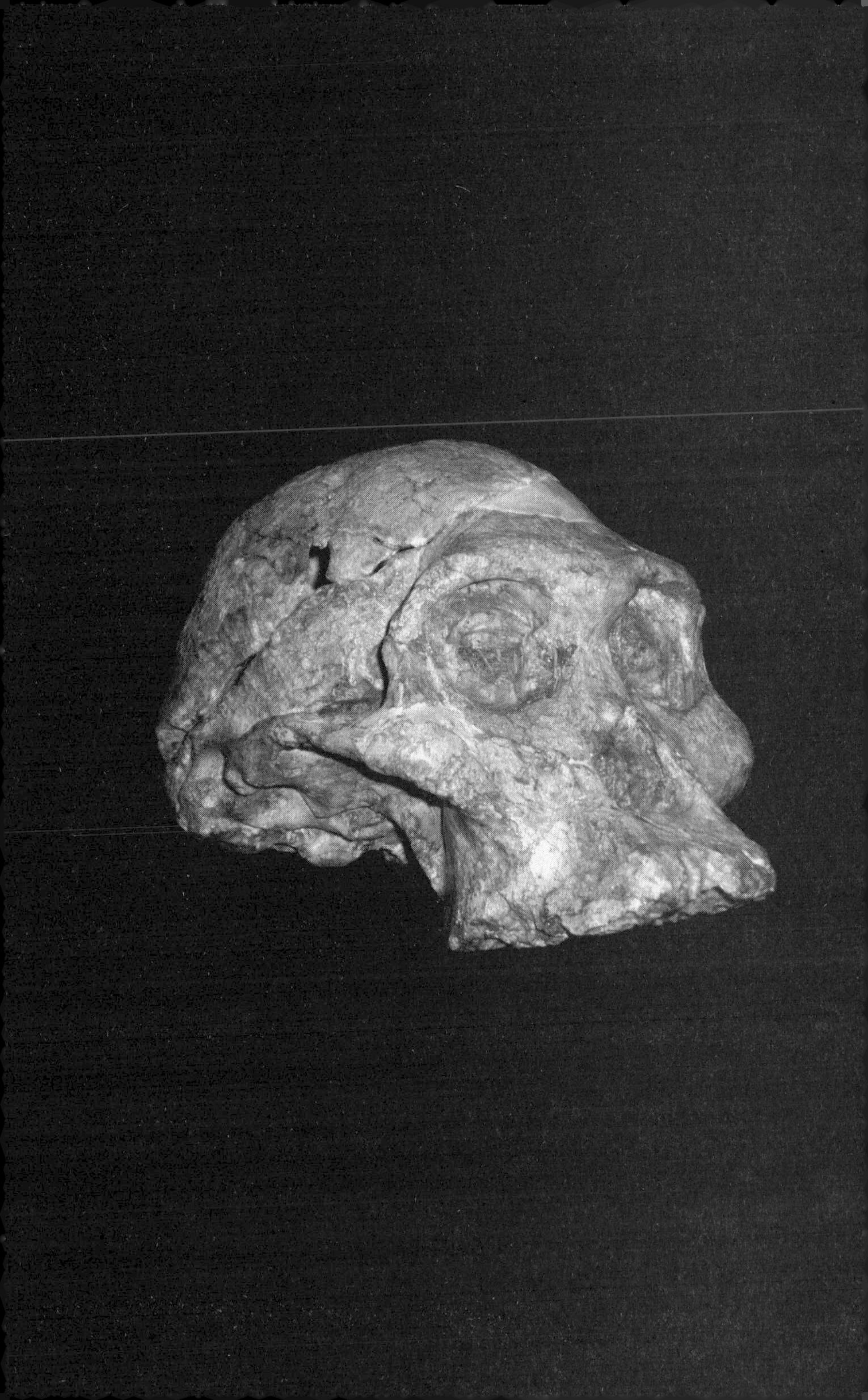

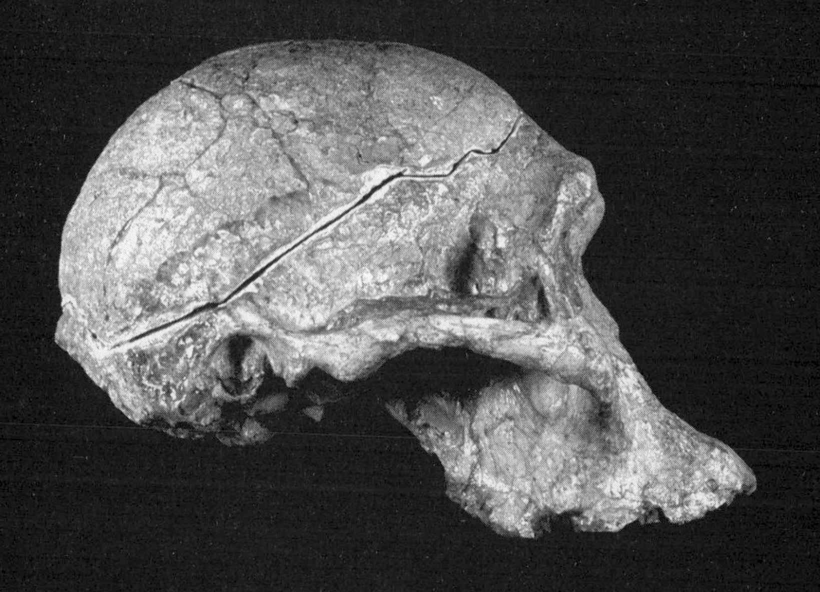

Two views of fossil skull of *Australopithecus africanus*, discovered in Sterkfontein, South Africa. (© Milford Wolpoff)

CHAPTER 3

Who Were the First Hominin Ancestors?

Who were our first hominin ancestors? What did they look like? And when did they appear? We often think that there is one correct answer to each of these questions, but just as with many other questions in human evolution, each answer depends on our ideas about our earliest ancestors. Paleoanthropologists generally agree that our earliest ancestors appeared about 5–7 million years ago in Africa.

Three fossil species discovered in the first decade of the twenty-first century, dated to 5–7 million years ago, are currently battling it out for the title of the earliest hominin ancestor. Since they are from a time period earlier than any of the hominin species discovered so far, any one of the three might be the one—if they are indeed hominin species at all. That debate is ongoing. These three species are not the only candidates vying for the title of first ancestor. Three other species, discovered in the twentieth century, are also still in the running. These species lived between 4.2 and 3 million years ago.

So, will the real first ancestor please stand up?

Did our ancestors have big brains?

In order to tell whether a fossil species is a hominin ancestor, we must first agree on what our earliest ancestor might have looked like. Charles Darwin talked about four characteristics that are unique to humans: big brains, small teeth, walking upright, and using tools. Although Darwin's model is no longer accepted literally, it has had a big influence on how paleoanthropologists hypothesize and model early-hominin origins. First, hominin ancestors were hypothesized to have some combination of these four characteristics, which in turn provided the explanation of how the hominin lineage started.

That humans have bigger brains than other animals relative to their body size is indeed one of the most striking characteristics of our species. In fact, our brains are large even in terms of absolute size. Thanks to these big brains, humans have the cognitive ability to process a tremendously large volume of information. Even our species name gives clues to our unique intellect: *Homo sapiens* means "sapient human," referring to the intelligence or wisdom of humans.

For a long time, scholars predicted that big brains should be the telling mark of ancestral humans. Other characteristics were thought of as only secondary to the development of our large heads. "Piltdown Man," discovered near London in 1912, fit this prediction quite well; it still had the ferocious canines of our ape predecessors, but it had a big brain. Piltdown Man became the pride of the English as our earliest human ancestor. Until 1953, that is, when it was revealed that the fossil specimen was a skillfully created farce, artificially made by combining the skull of a modern human with the teeth and mandible of an ape. It was a hoax.

As more fossil discoveries were made throughout the 1950s and 1960s, the predominant idea emerged that ancestral humans must have originated nearly 10 million years ago. All attention was on *Proconsul* and *Ramapithecus*, the fossilized remains of apes dating from this time period. They had a straight forehead and soft, gracile supraorbital bones (brow ridges). Paleoanthropologists noticed these human-like features and thought that perhaps these apes would finally give us insight into the earliest human ancestors.

Did human feet lead to human brains?

In 1967, Vince Sarich and Allan Wilson, both at UC Berkeley, published a short paper that would completely change the consensus about hominin origins. Their discovery was made not at an excavation site, but in an indoor biology laboratory. Research in biochemistry and genetics had suggested that human and gorilla lineages split 8 million years ago, and human and chimpanzee lineages split only about 5 million years ago. The fossilized apes that had once been thought to be the first hominins, *Proconsul* or *Ramapithecus,* were not the first ancestors or even close relatives, but distantly related apes from 10 million years ago.

But there was a problem with this DNA-based hypothesis: there were no fossil data to support it. Until the 1970s, the oldest hominin fossil remains were of the species *Australopithecus africanus,* first discovered in the 1920s in South Africa. But the *Australopithecus africanus* remains were dated to only 2–3 million years ago, which was too recent for it to be our earliest ancestor.

Starting in 1973, though, a handful of paleoanthropologists, such as Mary Leakey and Donald Johanson, made a series of huge breakthroughs. Important hominin fossil discoveries were made in East Africa, including at Hadar in Ethiopia and at Laetoli in Tanzania. These fossils were dated using radiometric methods and found to be 3–3.5 million years old. The new hominin fossil species was *Australopithecus afarensis.* The most famous of its kind found by Johanson and his research team was named "Lucy," and it was the oldest hominin ancestor fossil species discovered until that point.

The belief that it was the oldest hominin fossil species, however, wasn't the only reason why the discovery of *Australopithecus afarensis* marked a historic moment in human-evolution research. It was this species that definitively showed that humanity walked on two feet long before developing an enlarged brain. The brain of *Australopithecus afarensis* is barely as big as that of an adult chimpanzee. The teeth are larger compared to those of modern humans, and there is no evidence of tool use. In every way that we can see, *Australopithecus afarensis* looks more like an ancestral chimpanzee than an ancestral human—except for one thing: the fact that it walked on two feet.

The *Australopithecus afarensis* skeleton showed signs of upright walking, such as the shape of the pelvis, femur, and knee joint. And the existence of a "double arch" of the feet at the footprint site at Laetoli, Tanzania, is unquestionable evidence of bipedalism. A double arch is unique in human feet: one arch runs front to back, and another arch runs side to side—both working to cushion the shock of the body weight at contact. The discovery of *Australopithecus afarensis* precipitated a paradigmatic shift in the search for the earliest hominins. Not brains, but bipedalism, became the defining characteristic for an ancestral human. And other human characters would appear later in human evolutionary history. By this standard, *Australopithecus afarensis* held the title of the earliest hominin for a long time.

But *Australopithecus afarensis*'s position of glory was not to last. Starting in the mid-1990s, several ancestral hominins much older than *Australopithecus afarensis* were discovered, and all of them were also bipedal. *Australopithecus anamensis*, from 3.9–4.2 million years ago, is a great example. There is a heated debate about whether to consider *Australopithecus anamensis* as a third candidate (joining *Australopithecus afarensis* and *Australopithecus africanus*) for the earliest hominin ancestor. An obvious sign of bipedalism can be seen in the knee joint of the fossil specimen, and several traits in teeth, humerus (arm bone), and tibia (leg bone) are similar to those of *Australopithecus afarensis*—to the point that many suspect *Australopithecus anamensis* might be just another *Australopithecus afarensis*. Whether *Australopithecus anamensis* will be retained as a separate species classification, despite all the similarities with *Australopithecus afarensis*, remains to be seen.

New candidates appear

Things got more complicated in the early 2000s. Three new candidates, all older than *Australopithecus afarensis* and *Australopithecus anamensis*, joined the race for our earliest ancestor. Will one of these shed light on the dawn of humanity?

Two of these new candidates were found in 1999, right before the beginning of the twenty-first century. The first is *Sahelanthropus tchadensis*, discovered in Toumaï, Chad (central Africa). Judging by the fossil remains, this species is thought to have lived 6–7 million years ago. Considering that the majority of early-hominin fossils had been found in eastern or southern Africa, this discovery from central Africa was quite exceptional. The specimen, however, is represented by only pieces of a cranium, teeth, and a jaw, extremely fragmentary and distorted, making it problematic to draw firm conclusions about it. Early hominins often look similar to apes, except for traits associated with bipedalism. With only cranial fragments remaining, we cannot definitively tell whether *Sahelanthropus* walked upright, and hence we cannot be certain whether this is a hominin or an ancestral ape fossil. In fact, some paleoanthropologists argue that the skull of *Sahelanthropus* is closer to the gorilla lineage than to that of hominins.

The second candidate is *Orrorin tugenensis*, discovered in the Tugen Hills region of Kenya in East Africa. This fossil species also dates from 6–7 million years ago. The *Orrorin* femur (thighbone) shows traits of bipedalism; hence, there is a strong possibility that this is the earliest hominin fossil species.

If *Sahelanthropus* or *Orrorin* was indeed a hominin, then our beginnings would date as far back as 6–7 million years ago. There is, however, a possibility that these species were common ancestors to both humans and chimpanzees before the two lineages diverged. It is also possible that these species belong to yet another ape lineage. If so, then hominin beginnings might be more recent. With so few fossil specimens, we have not yet been able to unravel this mystery.

The third, and most recent, candidate for our original ancestor is *Ardipithecus ramidus*, discovered in Aramis, Ethiopia. This fossil species dates to 4.4 million years ago—later than *Sahelanthropus* or *Orrorin*, but earlier than *Australopithecus afarensis* or *Australopithecus anamensis*. Articles about *Ardipithecus ramidus* were published in 2009 in a special issue of the journal *Science*, and its discovery was titled the "Breakthrough of the Year." It made a big splash in anthropology, science, and society in general.

Another reversal: bipedalism in doubt

Why did *Ardipithecus ramidus* make such a big wave? It had long arms, big hands, and a big toe that diverged sideways like a thumb. This big toe (which, confusingly, wasn't the biggest toe on its foot) was the big problem. Such feet are usually found in tree-climbing apes, not in obligate bipeds that can move only by walking upright. If *Ardipithecus ramidus* was an obligate biped like us, the big toe should have been the largest toe on the foot and should have been parallel to the other toes, pointing forward, much as our big toes do. But the big toe of *Ardipithecus ramidus* shows that the species not only walked upright, but also may have been adapted to climbing trees. The consensus idea that the earliest hominins only walked upright was in doubt.

The ecology of the environment that *Ardipithecus ramidus* occupied presents another problem. For a long time, researchers theorized that the reason hominins became obligatory bipeds was the gradual disappearance of forests in Africa approximately 5 million years ago. The hypothesis explains that in West Africa, where there are still forests, the adaptations of forest-dwelling apes continue to this day in the form of chimpanzees and gorillas, while in East Africa, with its mixture of forests and grasslands, bipedal apes that could be active in the flat grassland survived and evolved as hominins. But the environment in which *Ardipithecus ramidus* lived was not grassland; it was forested. With the discovery of this species, then, the hypothesis that bipedalism evolved as an adaptation to flat grasslands may have started to crumble.

Of course, *Ardipithecus ramidus* may not be the earliest hominin species. *Sahelanthropus tchadensis* and *Orrorin tugenensis* are strong candidates as well, but they also have too many anomalies to be undoubtedly declared the earliest hominin. All three candidates could be members of the various ape lineages that roamed before hominins began, instead of being the earliest member of the hominin lineage. For example, they could be a part of the common ancestry between humans and chimpanzees, rather than being the first human ancestors *after* the divergence between the human and chimpanzee lineages. That would explain the apelike traits observed in *Ardipithecus ramidus* and other species, and while these candidates would be promising targets for the bridge between apes and hominins, they wouldn't be properly considered the earliest hominin. If this is the case, then the earliest hominin would belong to the genus *Australopithecus*—either *afarensis* or *anamensis*—and the date would be 3–4 million years ago.

Who were the first hominins? What did they look like? These questions have been explored for more than 150 years, since Darwin's evolutionary theory. The debate is ongoing, continually generating diametrically opposed arguments. New discoveries and new research can narrow our list of candidates for the earliest hominin, or they can unveil entirely new candidates, at any moment. As evolutionary theory

evolves, so does the research on hominin origins, through our countless questions and the search for answers.

EXTRA

WERE THE FIRST HOMININS TOOLMAKERS?

Along with big brains, small teeth, and bipedalism, making and using tools has been a hallmark of humanity. Louis and Mary Leakey, a famous team of paleoanthropologists who are responsible for many discoveries (see Chapter 15 for their story), named their discovered hominin—the first species of the genus *Homo*—"*Homo habilis*," which means "handy human." The implication of this first species name is that tools make our genus *Homo* unique. In that case, did the earliest hominins make and use tools? Maybe not. Artificially made stone tools start to be discovered from 3 to 2.5 million years ago, much later than the first appearance of hominins.

The first hominins had a brain size close to that of an average chimpanzee or gorilla: about 350–400 cubic centimeters (cc), which is quite small for the human lineage. It is unclear whether any species with such relatively small brains could make and use tools. Considering that chimpanzees are capable of moderately sophisticated tool use, we cannot say for sure that the earliest hominins did not use tools, or that archaeological signs of such tool use are undiscoverable. Not all tools will be discovered, especially if they are made of materials other than stones. Stone tools, however, are preserved for a long time.

A great example comes from *Australopithecus garhi*, discovered in 1996 in Ethiopia. *Australopithecus garhi* dates from 2.5 million years ago and was discovered with stone tool technology similar to the Oldowan industry (2.5–1.7 million years ago). Stone tools are defined by the way they are manufactured. Oldowan tools are made by striking a rock with another rock, discarding the flakes, and using the core with sharpened edges. The fossil hominin species *Australopithecus garhi* was also discovered alongside animal bones with clear cut marks made from stone tools. The *Australopithecus garhi* site has the earliest stone tools and the earliest evidence of stone tool use that we have discovered thus far. But the more shocking discovery was that the brain size of *Australopithecus garhi* was barely 450 cc. This is similar to the brain size of chimpanzees or other australopithecine hominins, and it shows that a big brain is not necessary for making and using tools.

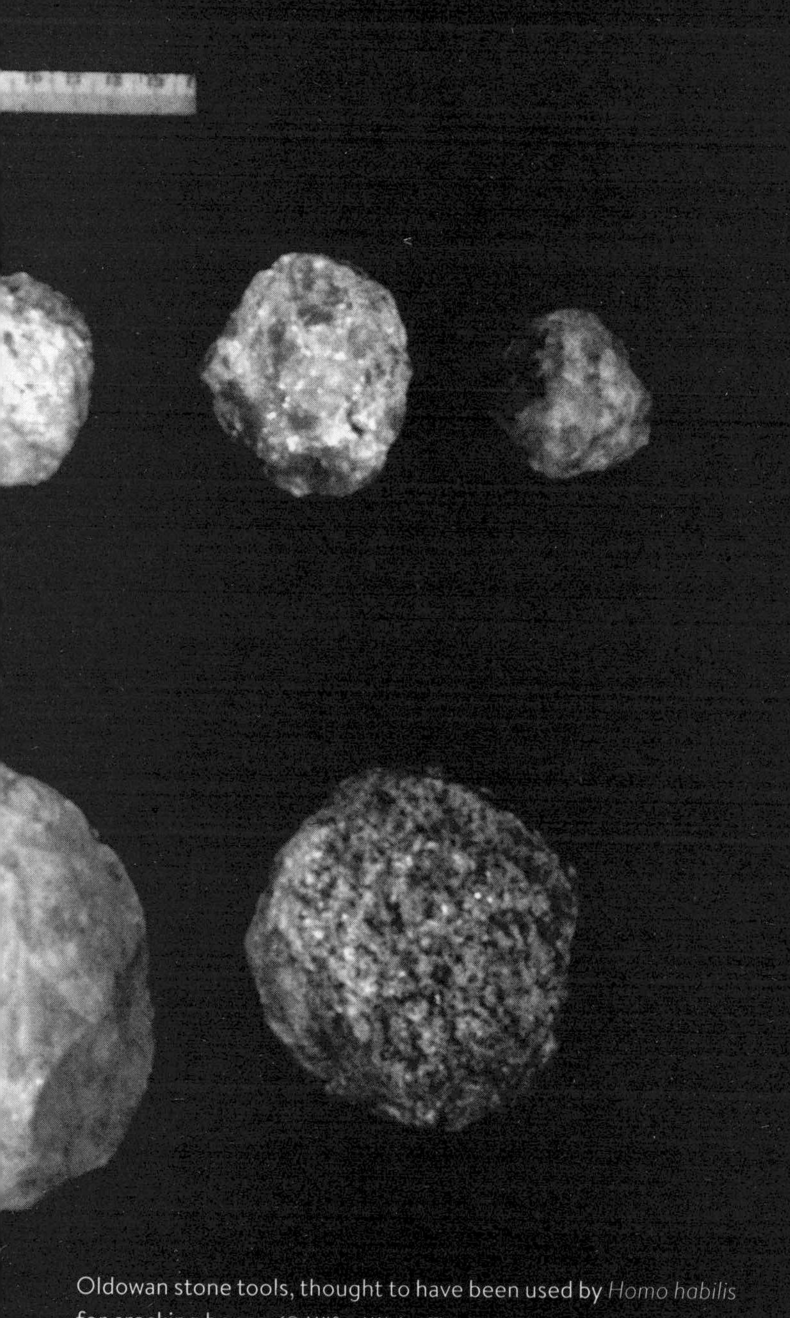

Oldowan stone tools, thought to have been used by *Homo habilis* for cracking bones. (© Milford Wolpoff)

CHAPTER 4

Big-Brained Babies Give Moms Big Grief

"Mother's Heart" is a Korean song often sung in May around Mother's Day, and it includes the line "You forgot all about the pain when you gave birth." Childbirth is accompanied by the greatest and longest pain many women will experience in their lifetime. Before the advance of modern medicine, childbirth was the biggest risk to the life of a young woman. Many died during childbirth. Many others died after a difficult childbirth, from hemorrhage or infection.

Even with the advent of significant advances in modern medicine and modern hospitals, mothers from one or two generations ago might still gaze at their shoes before lying down on the delivery bed and wonder, "Will I live to wear those shoes again?" Fortunately, today it is much rarer for women in modernized societies to die from childbirth, and with painkillers delivered directly into the spinal column, women can also give birth without feeling the full severity of their labor pains. Yet childbirth can still trigger fear and worry for women.

This is strange, since from an evolutionary perspective, having many babies is a marker of a successful life. Across most of the animal kingdom, life-forms grow through childhood and juvenility, and they begin to reproduce after maturing into adulthood. Having survived numerous dangers to life during vulnerable periods, a healthy adult will engage in reproduction as part of a natural life cycle. Perhaps that is why, for many animals, childbirth is not a big deal. Humans are the exception; for us, childbirth is a big deal and presents a serious danger. How did it get to be this way?

Bigger brain, smaller canal

Nonhuman newborns generally have a head that is not bigger than the birth canal of their mothers. This means that delivering offspring through the birth canal is not too difficult a task. Human newborns, however, are quite a different story. The process of being born is made much more difficult by our enlarged brain having to pass through a small pelvic canal.

Let's take a closer look. Early hominins from 4–5 million years ago looked very much like apes in terms of morphology. Their brain size was similar to that of chimpanzees: about 450 cubic centimeters (cc). The only difference was that early hominins walked upright. As time went on, hominin brain size continued to increase: 2 million years ago the brain doubled its size, to 900 cc; Neanderthal brain size surpassed that of modern humans; and fossil specimens from 10,000 years ago show a brain size similar to ours, at 1,400 cc. Our overall body size, however, has not changed much over the past 2 million years.

As the brain continued to increase in size while body size stayed relatively the same, several problems arose. To give birth to a big-brained baby, you want as wide a pelvis as possible. A wide pelvis means a wide birth canal. For efficient bipedal locomotion, however, it is better to have a narrow pelvis. When walking, the legs have to move or swing back and forth, from front to back. Legs that are wide apart produce a waddle effect, which is not energy efficient.

Humans had to make a choice between bipedalism and childbirth, and it appears we chose bipedalism; our pelvises have not really increased in size through the last 2 million years of human evolution, while our brain size has increased from 900 to 1,400 cc, on average. As a consequence, mothers have the unenviable task of forcing a big-brained baby through a narrow birth canal. The price for our upright mobility is the most difficult birth process in the animal kingdom.

Given this difficulty, it is amazing that human childbirth has been occurring successfully for millennia. To give birth to a baby with a brain bigger than the birth canal, the woman's pelvis has to soften at its joints to accommodate the extra width. The separating and widening of joints is triggered by a hormone called relaxin, and this softening happens throughout all the joints in a woman's body, not just the pelvis. Even with the widened joints, though, the birth canal is still smaller than the baby's brain, as evidenced by the still considerable risk and pain during childbirth.

After childbirth, women's joints mostly return to their original state, but they never recover completely. We often say that clothes do not fit the way they used to after childbirth. Even if your weight returns to its previous state, the shape of your body is forever changed. Moreover, the pelvis of a woman who gives birth to multiple children will show scars from repeated widening and narrowing.

Childbirth is a traumatic experience for human babies too. Before we turn to the one-of-a-kind experience for human babies at birth, let's first consider a primate baby's experience. A female monkey squats when she is giving birth, to get help from gravity. The fetus, when it enters the birth canal, faces the mother's navel. Therefore, when the baby is born it naturally faces the mother's face. The female, still squatting, stretches out her arms, helps the newborn out of her body, and holds the baby when it has cleared the birth canal. The baby monkey, with help from the mother, looks at its mom's face for the first time while being held in her arms, and eats its first meal.

Human childbirth: you can't do it alone

Human childbirth is much different from other primate births; in fact, it has literally made a 180-degree turn. During human childbirth, the uterus contracts and pushes the baby forward from behind, but the birth canal in front of the baby is too narrow relative to the baby's head size, so the fetus must also push forward with all its might. As the fetus pushes forward, it rotates its body to fit the narrow birth canal.

Karen Rosenberg, at the University of Delaware, and Wenda Trevathan, at New Mexico State University, have done much of the groundbreaking research into human childbirth. Their studies have shown that the fetus enters the birth canal headfirst. As with other primates, the fetus usually enters the birth canal facing the mother's front. Because the birth canal is oval, it has a long axis. And because the canal is so narrow, at some point on its journey the fetus must rotate so that the long axis of the birth canal is aligned with the long axis of its shoulders. A little farther down, the birth canal changes shape again; this time the long axis of the oval birth canal at the midpoint is perpendicular to the long axis of the birth canal at the starting point. The fetus rotates its body once more to align the long axis of its head with the long axis of the birth canal.

Hence, at birth the fetus no longer faces the mother; the baby has turned 180 degrees to face the mother's back. The human mother thus cannot make eye contact with her baby at birth, nor can she stretch her arms and pull the newborn out herself as other primate mothers can. A moment of wrong calculation, and the newborn's neck would be snapped backward. The human newborn needs to be received by someone else, who then hands it over to the new mom. In essence, someone else must be present to assist the new mother during human childbirth.

When a female animal feels the pangs of labor, she will often go to a safe and quiet place on her own. Generally, it is a place that she prepared for that very moment, such as an underground den. If someone else approaches her while she's in labor, the female might be alarmed and even kill the newborn; solitude is absolutely necessary for mothers in the animal kingdom. Human women in labor, however, do not want to be alone. In fact, if a woman in labor is left by herself, the stress hormone cortisol is secreted, prohibiting further labor, and sometimes childbirth comes to a halt. A woman in labor almost always has to be with someone whom she trusts and can depend on. Some of us may have heard stories from our mothers about going into labor while working the fields, delivering their baby all alone, and then resuming their work some hours later. This is definitely a story worth telling, because it is so rare!

In the long path of human history, before the rise of the modern hospital system, those who attended the birth of a child were usually the woman's mother, sister, grown-up daughter, or another woman from the same kin who had ample experience in childbirth herself. These trusted women stayed with the expecting mother, guiding her through labor and receiving the newborn, holding it so that its neck was protected, and then handing it over to the new mother. These women kin would also receive and take care of the placenta (afterbirth) that followed the newborn, and even help take care of many other daily affairs going forward while the new mother was bonding with her child. The very nature of human childbirth further confirms the evolutionary hypothesis that humans have needed others from the very beginning of our lineage: we are social animals from the moment of birth.

The true beginnings of humanity

When did this "social" childbirth, requiring someone to be with the woman in labor, start in human evolutionary history? In theory, we could gather information from fossilized female pelvises and newborns in order to answer this question: if the newborn head is too large for the pelvis and must have required the double rotation during childbirth, we may infer that childbirth must have been social. Such specimens are extremely difficult to come by, though, since newborns are rarely preserved as fossils. And the few fossilized pelvises of hominins are mostly males for some reason, making it quite difficult to get information about childbirth. ("Lucy," the female *Australopithecus afarensis*, is an extremely rare case.)

In 2008, Marcia Ponce de León and Christoph Zollikofer of the University of Zurich published some valuable new research on this topic. By scanning a rare Neanderthal newborn skull using computed tomography (CT) technology, they had found that Neanderthal newborns had to rotate twice for the big head to go through the narrow birth canal. Neanderthals also had to go through a very difficult and painful childbirth. This means that the origin of social childbirth goes as far back as 50,000 years at least.

Neanderthals were not the first hominin ancestors born with big brains. Also in 2008, a paper about a female pelvis from a *Homo erectus* specimen (a species that lived before the Neanderthals) was published in the journal *Science*. The shape of this pelvis, discovered in Gona, Ethiopia, was strikingly similar to that of a modern female pelvis. The shape of the birth canal, as reconstructed from the shape of the pelvis, was equally wide (side to side) as it was long (front to back), and it was quite different from that of *Australopithecus afarensis* ("Lucy"), our ancestral relative who lived earlier. Lucy's pelvis was short from front to back while wide from side to side, with a flat and small birth canal. The research team concluded from these comparisons that giving birth to big-brained babies most likely began with *Homo erectus*, approximately 2 million years ago.

Many traits are unique to humans. Some have become the qualifications we look for in fossils to confirm that they are indeed hominins.

Perhaps it is now time to add "belonging to a society from the moment of birth" to that list of uniquely human traits. The big brain of humans is the true hallmark of humanity, not because it signifies high intelligence, but because it made extreme sociality a prerequisite just to be born. By this thinking, *Homo erectus* is the first human.

EXTRA
CHILDBIRTH WITH FAMILY

Since the relatively recent advent of hospital delivery, childbirth is now no longer part of normal daily life. In modern society, even a normal delivery, without C-section, still usually happens in a hospital. Yet hospital delivery fully outfitted with modern hygiene measures often goes against the direction of our evolutionary history. For example, women undergoing labor usually have an easier time if they sit upright; that way the baby is born in the direction of gravity. And, as mentioned earlier, when labor starts, women need to be with someone they trust and can depend on. In hospital births, however, women are often lying flat and give birth only after all family members have left the room and those who remain are a rotating medical staff consisting of unknown but medically trained men (and recently, women). Lacking the comforting presence of familiar people, a woman in labor often becomes tense and anxious, which may cause contractions to cease and eventually necessitate an emergency C-section.

In response to this counter-evolutionary environment, more and more hospitals are having women go through labor sitting up, and allowing them to experience childbirth and recovery in a family delivery room with their family members in attendance. In addition, the number of women opting for home births with their family members and for the assistance of a midwife is on the rise as well. This is a positive trend that fosters our evolutionary need to be among kin at our most vulnerable moments.

CHAPTER 5

Meat Lovers R Us

Imagine a human child, perhaps four or five years old, chasing down gazelles running away at top speed—outrunning even lions—in an African landscape. Is this possible? Of course not. But if early hominins hunted for meat in the way we imagine they did, this is the scenario we are forced to include in our evolutionary history. Indeed, early hominins were only about as tall as a human preschooler, with hunting skills barely developed.

Yet it's clear that humans love to eat meat. If eating meat was considered a skill, we would be masters at it. This skill appeared in the middle of the human evolutionary history, about 2.3 million years ago with the start of the genus *Homo*, but the manner in which we first started to obtain meat—or animal fat and protein, to be precise—was not hunting, as we tend to think. We like to imagine a cartoonlike scenario in which a caveman chases after game with a spear or a stone ax, but such pursuit became possible only after substantial time had passed in our evolutionary history. Spears appear only within the last 30,000 years; stone hand axes, about 2.5 million years ago. But if we are such meat lovers, how exactly did we get our meaty fix in the beginning? Before we look at that question, let's explore how we were even able to digest so much meat in the first place.

A new primate with old taste buds

Humans are primates. Our first primate ancestors originated between 80 and 65 million years ago, living in the treetops with fruit and leaves as their main diet. These first primates could fit inside the palm of your hand (think tarsiers) and therefore required minimal caloric intake for sustenance, most likely from leaves and fruits. This diet is a contrast to that of modern small-bodied monkeys (yet much larger than tarsiers), which get their protein and fat nourishment from insects and larvae, but it is similar to the diet of apes with larger body sizes, such as orangutans and gorillas, which are almost exclusively herbivores.

For these larger apes, vegetation is likely the only option, since there is no guarantee of being able to find the amount of animal meat needed to sustain such a large body. Chimpanzees, the most closely related lineage to humans, acquire and eat meat—in groups they hunt baby baboons, and they use twig tools to dig out and eat termites—but animal-based food makes up a negligible proportion of their overall diet, especially compared to the amount of meat that humans consume.

The fact that apes, our closest ancestors, are mostly herbivorous implies that our earliest human ancestors were also likely herbivorous. In fact, paleoanthropologists think that early hominins from 4–5 million years ago had a diet based on plants, just like the other apes did. The shape of the teeth and the deep mandibles that belonged to our early hominin ancestors from around this time hint at a substantial amount of mastication (chewing), which would have been necessary to process large amounts of vegetation. (Plant-based foods have a lower caloric density than do animal protein and fat, so our ancestors would have needed to consume a lot of plants to sustain themselves.)

In addition, the first hominins had a brain size similar to that of modern chimpanzees, suggesting they lacked the strategic processing power to hunt moving prey or to track and scavenge from other hunters. Herbivores don't need large brains, because plants don't move. These morphological traits—our ancestors' chewing-oriented tooth shape, deep mandibles, and small brain size—are usually seen in herbivores rather than carnivores, and taken together, the evidence seems to suggest that the first hominins did not rely much on animal-based food.

Brave hunters? More like carcass scavengers

Most animals love meat. Carnivores, of course, can eat only meat; perhaps surprisingly, herbivores and omnivores also like to eat animal fat and protein. Likewise, humans love meat—perhaps more than any other omnivore on the planet does. But we had to overcome many challenges to be able to consume our food of choice.

In 1974, a strange *Homo erectus* fossil was found at Koobi Fora, the famous paleoanthropological site in Kenya. Named "KNM-ER 1808," the fossil was estimated to be 1.7 million years old, using the radiometric dating method. Scientists noticed something odd about this fossil's bone structure: the cross section of the bone was extremely thick. Paleoanthropologists hypothesized that this early *Homo* (the group of hominin species that includes *Homo habilis, Homo rudolfensis, Homo erectus,* and *Homo ergaster*) likely experienced bleeding in the bone around the time of death. In the same way that an inflamed site swells up, bleeding in the bone causes it to become thick. The most plausible culprit for this bleeding was hypervitaminosis of vitamin A—that is, an overabundance of vitamin A in the body.

That was strange. Hypervitaminosis of vitamin A is a likely result from overeating the innards, especially the liver, of other carnivorous animals. "I guess our ancestors ate a lot of meat, no big deal" would be an easy response, but on closer consideration, there's a mystery here. As the fallen *Homo erectus* indicated, the evolving human body was clearly not well adapted for eating large quantities of meat. How is it, then, that hominins who were most likely reliant on plant-based food became such meat lovers that they could eat enough animal protein to die from hypervitaminosis A? The dramatic change in hominin diet may have been related to the dramatic change in the environment.

Africa became dry during the Pleistocene, which lasted from approximately 2.6 million years ago until about 12,000 years ago. Woodlands dwindled and grasslands began to expand. Competition became increasingly fierce for plant-based food. All of this environmental change did not favor hominins, who relied on a plant-based diet. To make matters worse for hominins, the remaining woodlands were monopolized by the hominin predecessors *Paranthropus* (also called *Australopithecus* by some scholars), who were only one-fourth the size of modern gorillas but had strong jaws and teeth as big as those of modern gorillas. *Paranthropus* could eat a lot of plant-based foods, including bark and roots, and so could survive these hard times. Early *Homo* adults, on the other hand, were about 3 feet tall (the size of an average four- to five-year-old modern human child) and had smaller teeth. Eating bark and other tough plant-based foods and hunting game were both out of the question. That left really only one option: early *Homo* had to survive by scavenging for animal fat and meat.

If it's difficult to hunt live game, why not just eat the leftover carcass? Lions get their "lion's share" from the intestines of their fresh kills, then take off for a nap to digest the meat. The fresh kill is left intact, except for the innards. So in theory, early hominins could have feasted on the remaining meat. In practice, even scavenging does not come easily. After the lions leave, other scavengers, such as vultures and hyenas, descend upon the leftovers. A vulture can stand as tall as 3 feet—eye to eye with an early hominin. Furthermore, these birds always travel in groups. Early hominins could not compete with these first-round scavengers for the leftover meat.

Instead, early hominins had to devise an innovative strategy to help them obtain needed calories. In fact, it's not really much of an innovation: they simply waited until all competitive players, from lions to vultures and hyenas, were done with their meals and had left the scene. There was only one thing left behind at that point: the bones. Very few predators can break through bone, but bones can be a source of rich nutrition; the limb and cranial bones contain precious marrow and brain matter. Pure fat.

There is a reason, however, that all that nutrition usually remains locked away inside bones. Bones are very hard. Limb bones in particular can be hard enough to be used as weapons, and early-hominin teeth certainly could not crack them open. Instead, our early ancestors found a way to crack these bones open and get to the marrow by using stone tools. We refer to these early crude stone choppers as Oldowan tools, thought to be made by *Homo habilis* and/or *Homo rudolfensis*. It is no exaggeration to say that modern human civilization owes its existence to these little pieces of stone.

The consequences of meat eating

Humans started to eat leftover marrow to survive as the woodlands gave way to ever-expanding savanna. During this transition, surprising things began to happen in our evolution. The intake of high-calorie food led to an increase in cranial capacity. The brain is an organ that is costly to make and costly to maintain (in terms of caloric energy). To have a big brain, you must secure a calorie-dense, high-quality food source. As we added meat to our diet out of necessity, we also made it possible to increase our brain size.

Regular consumption of high-fat, high-protein food also led to increased body size. The first hominins, from 4–5 million years ago, had a brain size similar to that of adult modern chimpanzees: 400–500 cubic centimeters (cc). The brain of *Homo habilis,* 2–3 million years later, had increased in size to as much as 750 cc. But body size still hovered around 3 feet. Half a million years later, *Homo erectus* appeared. *Homo erectus* had a brain size as big as 1,000 cc, and a body size as big as almost 6 feet. A human ancestor with a big body and big brain had finally appeared on the stage.

Equipped with a big brain and a big body, hominins could at long last start to pursue living animals. This is when we started to resemble the stereotypical "caveman" hunter image; it's worth pointing out that we didn't get there until relatively late in our human evolutionary history. But soon after, hominins became adept hunters, thanks to their excellent innovation, physical strength, and stone tools.

Up to now in this discussion, I've said only that the first hominins had no choice but to start eating meat. One question remains unanswered: Just as gorillas and chimpanzees cannot digest a large amount of meat, even if they love meat or have nothing else to eat, early hominins likewise could not have digested a large amount of fatty, meaty food easily. How did they come to acquire this ability?

This last problem was solved through evolutionary forces of natural selection. Fatty food can be digested by means of a compound called apolipoprotein. Apolipoprotein functions like dish detergent for the body. It binds to a fatty compound, then leaves the blood vessels, clearing the blood of fat molecules. In particular, APOE4 (apolipoprotein epsilon-4) is especially efficient at lowering the fatty protein level in blood. This compound is a result of a genetic mutation that occurred about 1.5 million years ago, when *Homo erectus,* with its big brain and big body, began making Acheulean hand axes by working on both sides of the stone core.

The KNM-ER 1808 fossil remains capture an intriguing moment: hominins had begun to eat the liver and innards of other animals, triggering a dramatic change in human evolutionary history, yet the fact that this hominin died from bone bleeding implies that our ancestors were not yet fully able to digest a lot of fatty foods and animal protein. The place of the amazing KNM-ER 1808 lies in the middle of a tumultuous time in human evolution.

Humans completed the genetic adaptation toward successful meat eating and came out as survivors after a long and harrowing road. We could finally rely on acquiring animal protein and fat through hunting. But hunting skills alone were not enough to make us meat lovers; we needed genetic adaptation to be able to ingest so much meat.

EXTRA

MEAT EATING IN EXCHANGE FOR DEMENTIA?

Apolipoprotein, the protein that clears the blood of fatty compounds, is also associated with critical diseases often linked with Alzheimer's disease, dementia, and stroke. Some researchers think that the apolipoprotein gene is the direct cause for these other diseases associated with old age. If so, why do humans still have such a dangerous gene? Shouldn't natural selection have eliminated a gene that causes such serious diseases and death?

That we still carry around a gene that leads to the aging process is explained in evolutionary biology by various hypotheses, one of which is pleiotropy, the association of one gene with several traits. Let's assume that a particular pleiotropic gene is beneficial during childhood and youth, but is harmful during old age. Will this gene disappear from the gene pool because of the harm it bestows in old age? Since it is pleiotropic, this gene is responsible for both the benefit during childhood and the harm during old age. The benefit during childhood and youth outweighs the harm later in life, and thus the gene will not be weeded out by natural selection.

Natural selection favors childhood and youth more than the years after our reproductive capability. This tendency can be used to explain APOE4 (apolipoprotein epsilon-4). The gene for this protein is still in our gene pool because the benefit from clearing the blood of fatty compounds is greater than the harm associated with Alzheimer's or stroke. Our ability to eat meat comes with a cost. You might then ask, "Would becoming a vegetarian eliminate the risk of dementia in old age?" The answer, unfortunately, is no. The apolipoprotein gene helped us survive by transforming us into meat lovers, and it is here to stay, regardless of your diet choices today.

Stone tools and hippopotamus bones on an archaeological site in Olorgesailie, Kenya. (© Milford Wolpoff)

CHAPTER 6

Got Milk?

Milk stinks. I can distinctly remember furtively passing my share of milk to my brother; failing that, I drank as much as I could in one gulp holding my breath, so that I did not have to smell its offensive odor. My classmates would tease that I was dark and short because I did not drink milk. Indeed, it seemed to me that those friends who liked drinking milk were taller and had fairer skin. Often, I would see milk advertisements that promised those same physical improvements. In the United States, the "Got Milk?" advertising campaign was quite popular through the 1990s and early 2000s. This series of advertisements featured famous, beautiful people with a milk mustache, inviting the viewers to drink milk. But I never liked drinking milk.

I'm not alone; many people become sick from drinking milk or eating ice cream. Soft, cold, and sweet ice cream melting inside our mouths in the middle of a hot summer day can make us feel luxuriously happy. But those who cannot tolerate milk or ice cream may feel nauseated or gassy, or even experience diarrhea or vomiting. Milk makes them feel terrible. But why? Lactose, the carbohydrate found in milk, is difficult for the body to digest, requiring the production of a very special enzyme, called lactase. People who lack the ability to produce lactase and thus cannot digest milk find the experience excruciating.

In the United States, this condition is called "lactose intolerance" or "lactase deficiency," and it is considered a disease or a disorder. There is no cure for this condition. We cannot build tolerance for lactose by drinking a little milk every day, nor does it necessarily get worse when a lot of milk is consumed. Symptoms of lactose intolerance may be relieved through artificial means, such as by taking pills that contain the enzyme lactase or by drinking lactose-free milk. But should people be forced to drink milk even if it means taking additional supplements to help digest it?

Adults who can drink milk are strange

In the United States, lactase deficiency has long been considered an abnormality, even a disease. But as early as the 1970s, biological anthropologists and geneticists started to notice a worldwide pattern of lactase deficiency. They raised questions about this "disease" and argued that the abnormality lies with the adults who *can* drink milk, not with those who cannot.

In fact, lactase deficiency is not a disease. We, as mammals, are born with the active gene to make the enzyme lactase. After all, we have to drink our mother's milk! The gene for making lactase remains active throughout infancy, as milk is our only source of nutrition early in life. During the weaning period of late infancy, the gene becomes less active, and lactase is made less and less. After weaning, we begin to rely more on solid foods, and lactase is made even less, while other digestive enzymes are made more. Finally, in adulthood, the gene that makes lactase becomes inactive—or at least it should—and the milk sugar lactose can no longer be digested. Hence, lactose intolerance in adults is a condition that occurs naturally and normally when adulthood is reached.

Anthropologists, comparing different cultures in the world, have long argued that lactase deficiency is not a condition that deserves further study; instead, lactase *persistence* into adulthood is the condition that warrants special attention. A look at different populations all over the world tells us that anthropologists are correct. On average, the ratio of adults who can digest milk lingers between 1 and 10 percent of the whole adult population. Most of Asia, most of Africa, and much of Europe fall into this range, meaning that most human populations follow the trajectory of a normal mammal.

In only a few populations can the majority of adults digest milk. They live in Sweden and Denmark in northern Europe, Sudan in Africa, Jordan in the Middle East, and Afghanistan in South Asia. In these populations, the proportion of adults who can digest milk may be as high as 70–90 percent. In these regions, not being able to digest milk would indeed be abnormal. As for Americans, who are mostly immigrants, the proportion of adults who can digest milk depends on the place of origin before immigration. A high proportion, 70 percent or more, of African Americans, Asian Americans, Middle Eastern Americans, and Native Americans are lactose intolerant as adults, while a high proportion of Americans originally from northern Europe can digest milk as adults.

Adult milk drinkers are only 10,000 years old

The populations with lots of adult milk drinkers have one thing in common: they all have a long history of dairy economy and cattle husbandry. Pastoralists are known to drink milk and other dairy as the main staple of their diet. Researchers have long thought that the ability to digest milk into adulthood should be correlated with animal husbandry and a dairy economy.

"Of course you would be able to drink milk as an adult if you were from a population that drinks a lot of milk!" you might be thinking. But such a conclusion is only an idea—albeit a very convincing one—until it can be postulated as a hypothesis that can be tested with data. After the discovery of the mutation in the gene that codes for lactase, called LCT, however, that hypothetical correlation could finally be tested. As expected, the LCT mutation led to persistent production of the enzyme lactase into adulthood, enabling the adults with that mutation to drink milk. And populations with a high frequency of lactase persistence were those with a long history of a dairy economy and animal husbandry.

Perhaps that is not surprising at all. But don't get disappointed quite yet, because the full story is not so simple; in fact, it's a perfect example of how exciting scientific research in anthropology can be. Surprisingly, the condition of lactase production was not the result of a single-point mutation in the genes. For example, the lactase mutation in the DNA sequence in Sweden was different from the mutation in Sudan. This means that the reason these two populations have a high proportion of adult milk drinkers is not migration; the Swedish did not migrate to Africa, nor did the Sudanese migrate to Europe, spreading their specific mutation. Instead, selective pressures on each population produced different mutations in the lactase gene, both resulting in the persistent production of lactase into adulthood. What a coincidence!

It is not clear which occurred first—the mutation or the dairy economy within these populations. If the hypothesis that drinking lots of milk caused the frequency of lactase mutations to increase is true, then the dairy economy must have appeared *before* the mutations. Geneticists and anthropologists have collaborated to test this idea with data, to discover when these mutations might have occurred. In Europe, the dairy economy began after the Neolithic

Age, starting 9,000 years ago. If the European Neolithic people already had the mutation, people with a lactase mutation must have started the dairy economy. If the opposite is true, that the Neolithic people did not have the mutation, we can conclude that selective pressure from the dairy economy triggered the increase in frequency of the genetic mutation.

In 2007, researchers from the Gutenberg University in Germany and University College London in Great Britain succeeded in extracting ancient DNA from a Neolithic skeleton dated to before the advent of the dairy economy. Ancient DNA that is extracted from a fossil is extremely difficult to isolate because of the fragmentation and contamination of the sample. Yet thanks to recent advances in genetics, even such ancient genomic data can be extracted with high reliability. Analysis of the ancient DNA extracted from the Neolithic skeleton showed that it did not have the mutation to digest lactase. Therefore, our lactase mutations started to increase in frequency only within the last 10,000 years—*after* the start of the dairy economy.

Why did milk drinkers increase?

The data support the hypothesis that the lactase mutation was an adaptation to the dairy economy and pastoralism. This mutation, present only within the last 10,000 years, is an extremely recent development from the perspective of human evolutionary history, yet in some populations the mutation's frequency is already as high as 90 percent. Why has the recent selective pressure to digest milk been so strong? Such speedy dispersal of a mutation means that the people with that particular mutation have been leaving significantly more offspring in the next generation than those without the mutation. In other words, people who could not drink milk either died younger or were not as reproductively successful as people who could.

What could be so critically important about milk that it became a matter of life and death? Several hypotheses address this question. The first is that drinking milk makes people taller, which may have been a significant advantage. Indeed, northern Europeans are considered some of the tallest people in the world, and they drink a lot of milk. But it is not clear that they are tall *because* they drink milk. No study has clearly shown which compound in milk increases height and weight through which chemical process. Furthermore, it has not been clarified how being tall could be a selectively advantageous morphological trait.

Another argument is that milk provides valuable calcium and protein. But calcium and protein can also be acquired through cheese or yogurt, which are easier to digest than milk because of the way lactose is transformed through the fermentation process. Why would a human population wait around for a genetic mutation, when an easier and more immediate pathway of adaptation is possible through culture? In fact, many cultures in the Middle East have dairy economies, but the proportion of adults who can digest milk is lower than in northern Europe. It is possible that the lower frequency of milk-digesting enzymes in Middle Eastern populations is due to ingesting milk in a form that is easier on the system, such as cheese or yogurt.

One final hypothesis is that milk provides vitamin D, an important component for absorbing calcium in the body. It is also the only vitamin that we can synthesize on our own from sunlight. Since there is not much sunlight in northern Europe, the vitamin D hypothesis is convincing. This argument, however, is less convincing when we consider other regions where mutations for lactose digestion are also prevalent, such as Sudan.

In conclusion, the mystery continues. We still don't know why milk has had life-or-death importance on our development. Yet we can still enjoy milk and ice cream without understanding why.

Coevolution of humans and cows

In the last 10,000 years, humans evolved the ability to drink milk into adulthood from a series of genetic mutations in the lactase gene. But it is not just humans who underwent big evolutionary changes. Milk itself has changed, because the genetic makeup of the cows that produce the milk has changed through domestication. Just as the rice we eat today tastes quite different from wild rice, milk from domesticated cows is quite different from "wild" milk. Milk also underwent changes to fit the taste buds of humans, to be more like human breast milk. We have changed cow milk to fit our own tastes. Through animal husbandry and artificial selection, we have altered cows' genes to suit us. Some people argue that it's harmful to feed human babies cow milk, but perhaps it is the calves instead that should be protesting about the strange-tasting milk, fit to feed humans, that they have to drink.

From the 1960s until just a few years ago, scientists thought we had stopped evolving after the Neolithic Age, when agricultural civilization started. We thought our bodies were remnants of an era before 10,000 years ago. Instead, genetics and anthropology have shown that humans have continued to evolve recently, at an even faster rate than we did over the previous 5 million years. (I'll revisit this idea in Chapter 22.)

EXTRA
GOT MILK?

The "Got Milk?" advertising campaign featured famous stars such as Hugh Jackman from the X-Men movies and the popular singer Rihanna. These advertisements made it look like it was only natural to drink milk as adults. But, just as in the rest of the world, a lot of adults in the United States cannot digest milk.

From the sixteenth to eighteenth centuries, and then again in the mid-nineteenth century, waves of immigrants from northern Europe brought the dairy economy with them. As explained in this chapter, these people had the mutation to digest milk into adulthood. As they became identified with mainstream culture, drinking milk as adults also came to be considered normal. Not being able to digest milk became rather provincial. It was not too long ago, only in the late nineteenth to early twentieth centuries, that milk came to be considered an essential part of the American diet.

The interesting thing is that drinking milk is going global. Like McDonald's and Starbucks, drinking milk is perceived to be part of a "development" or "Westernization" package. In the 1990s, the two countries that showed the biggest increase in milk consumption were China and India, yet both China and India have a high proportion of adults who cannot digest milk. Considering that both of these cultures boast a diverse cuisine rich in all necessary nutrients, it is hard to believe that the motivation to drink milk as adults lies purely in the nutritional benefit of milk drinking.

CHAPTER 7

A Gene for Snow White

Every summer, a wave of new beauty products promise to make our skin lighter. Do they work? Can we make our skin white? Not really. We are born with our skin color, which is derived from a particular combination of genes coding for the skin pigment melanin. Although sunlight and exposure play a role in determining skin color, sunlight (or lack thereof) cannot completely change the skin color we are born with. This simple fact—the amount of melanin in our skin—has been the source of many sensitive and controversial debates throughout human history.

Surprisingly, though, it was only in the seventeenth century that skin color began to be used by Europeans to classify humans into different races of black, white, yellow, and red. As they circumnavigated the world, Europeans encountered people who looked dramatically different from themselves, with different skin color. Beginning in the 1960s, however, anthropologists started to question this "fact" of racial difference. Biological anthropologists such as C. Loring Brace and Frank Livingstone, both at the University of Michigan, argued at that time that skin color was not a fundamental and fixed characteristic that could be used to demarcate racial categories; as mentioned already, skin color can change to a degree, depending on exposure to ultraviolet sunlight. In an article published in 1962 with the title "On the Non-existence of Human Races," Livingstone left this famous quote: "There are no races, there are only clines."

If you look at a world map, you can see that there is some relationship between the strength of ultraviolet radiation in different regions and the skin color of the people who live in those regions. Those who live in regions with a lot of UV radiation tend to have darker skin, and those living in regions with less UV radiation tend to have lighter skin. And there is a continuum of skin color between black and white—a gradation of all the other supposed skin color categories. Skin color is nothing more than the result of adaptation to a specific environmental condition, not an inherent and essential trait that can be used to define racial categories. Furthermore, it has become well established that "race" is not a biologically meaningful category along any human trait; it's a social construct.

It looks as if there is no need for any more debate about skin color. Yet this conclusion is only the beginning of a series of complex questions about the mysterious nature of skin color.

Humans have peach fuzz

Mammals have fur that protects their bodies. Fur keeps dangerous things away, such as ultraviolet radiation, thorns, fangs, and other elements of nature. Fur maintains a stable body temperature by keeping the layer of air around the body warm or cool. Thanks to fur, mammals have the ability to survive in a wide range of environments, regardless of the temperature.

Given all the advantages of fur, human skin is very strange. We are covered not with fur, but with hair, a rare trait among mammals. The only other mammals without fur either have been selectively bred that way, or live in environments that lack sun exposure. Among mammals who live aboveground and are regularly exposed to sunlight, humans are uniquely naked.

Compared to other animals of similar body size, however, humans do have a similar number of pores and a similar number of hair follicles covering their bodies. We just appear naked because our hair is short and fine, akin to a light-colored "peach fuzz." Humans became naked not by losing their fur per se, but by changing the kind of fur they had.

When did this transformation happen, and why? The most convincing hypothesis has to do with our ability to eat large amounts of meat. Early hominins were probably fruit-eating vegetarians, but that changed about 2.5 million years ago. Although all they could get was scraps of meat and marrow left behind after other animals had taken their meals from a carcass, there was still enough meat to fuel an increase in brain and body size, ultimately leading to more sophisticated stone tool technology for hunting.

Mammals with fur usually hunt during the evening, in the early morning, or at night, when the temperature drops. Think of a lion on the African savanna, with its mane and lustrous fur. A male lion or lioness sure looks beautiful, but could he or she run around in the middle of a hot day? It would be like me trying to run around the savanna in a long fur coat. I would likely collapse from heatstroke, and for a lion it's much worse. Moreover, in the middle of a hot day, despite the shade, predators often open their mouths and pant to dissipate extra body heat, just like a dog does on a hot summer day. For furry mammals, even remaining perfectly still in the midday heat is exhausting; never mind running after a gazelle galloping away at forty miles per hour.

Hominins seized the opportunity and became hunters during the day when other predators were resting. But this change would have been impossible with a body covered with fur. How were early hominins able to shed their fur coats?

Birth of the naked human and dark skin

Suppose a mutation leading to extreme hair reduction and nakedness appeared accidentally. Hominins with this mutation would have been able to conquer the hot days in Africa by getting rid of extra body heat through the evaporation of sweat on their naked skin.

Not all traits are purely advantageous; going furless came with a cost. Once hominins began to regulate their body temperature through sweating, they became that much more dependent on adequate water sources. It must have been difficult to find drinking water in Africa during its dry seasons. Where and when to get water would have been critically important information, so it became essential for early hominins to store this seasonal knowledge in their memory and to communicate that knowledge to the next generation. Moreover, frequent visits to the same water hole are dangerous, so an ability to effectively communicate those dangers became important.

Ultraviolet radiation would also have been a problem for our furless ancestors. Without the fur that blocked it, hominin skin was directly exposed to UV radiation, which can cause severe burns and lead to infections. But more important from an evolutionary perspective, UV radiation can also induce birth defects by destroying folic acid in the body. Extensive sun exposure has the potential to reduce the number of viable offspring; this selective pressure would lead to an adaptation to block the sun's radiation.

In the human body, this function is performed by melanin, the chemical protein for skin pigment. Melanin is produced by specialized cells in the human body called melanocytes; the more melanin is produced, the darker the skin becomes. This is precisely why we think the first furless hominins in Africa had darker skin. After exchanging fur for sweat, ancestral humans had to adopt darker skin to survive. In contrast, animals with dark fur, like chimpanzees, tend to have lighter or white skin. Because fur blocks sunlight, there is no need for the underlying skin to be pigmented, and therefore no reason to have skin color.

Following this logic, the first humans with naked skin must have quickly developed darker skin, and all humans must belong to the "black race." But not everyone in the world has dark skin. In fact, some people's skin is almost white. So, how did so many "Snow Whites" come about in human history?

We got back lighter skin

As modern humans spread all over the world, they moved from equatorial regions to more northern regions with less sunlight. In particular, the time when hominins spread far and wide was during the repeated cycle of glacial and interglacial periods of the Ice Age. During the glacial periods, the frequency of cloudy days increased and there was not as much sunlight. The reduction in direct sunlight meant there was less UV radiation to block, so humans did not need as much melanin. By itself, however, this change should not have been enough to trigger whiter skin; not needing melanin is not the same as needing lighter skin. If melanin was completely irrelevant to survival, it would not matter if our skin was dark or pale.

It turns out that skin color is not a mere elective. It is critically important for nutrient regulation; in the case of folic acid, in fact, skin color could be a matter of life or death. In the same way that you need more melanin to survive in a region of strong sunlight, you also need less melanin to survive in a region of weak sunlight. Why? Because our bodies need some UV radiation to synthesize a very important vitamin: D. Vitamin D plays a critical role in calcium metabolism and nutrient absorption; without it our bones lose their rigidity and become deformed because calcium cannot be absorbed properly. If adults go without vitamin D for a prolonged period of time, or if growing children experience vitamin D deficiency during their growth periods, they can develop rickets or other harmful conditions.

A deformed bone does not necessarily mean death. For women of reproductive age, however, bone deformity can certainly present a life-or-death situation. Specifically, a deformity of the pelvis would have a devastating effect on a mother's childbirth prospects. In the face of such a clear reproductive threat for mothers and babies, humanity adapted by having lighter skin to better synthesize vitamin D in more northern regions.

The world map of skin color shows that skin color is gradually distributed, with the gradations aligned by latitude. Darker skin is found in the equatorial region, and lighter skin away from the equator, as predicted by the vitamin D hypothesis of skin color. The reason for the gradation is that annual solar radiation varies by distance from the equator, with more radiation around the equator and less radiation farther from the equator.

The genes for skin color were discovered only recently; the first such gene, one responsible for melanin production, was reported in 1999. Since then, at least twelve other genes involved in skin color have been discovered. Different combinations of these different genes result in the variation of skin color. Some genes are regulatory: they "turn on" or "turn off" the gene that makes melanin. Snow White, from the famous nineteenth-century German fairy tale, probably had such white skin because she had one of the genes that turn off the melanin-producing gene.

Even though the geographic distribution of dark and light skin tones is consistent by latitude, the frequency of each of these specific genes varies by continent. For example, while Polynesians (who live in the western Pacific) and equatorial Africans both have dark skin tones, those tones are different, on average, in color and brightness. Moreover, different genes are responsible for the lighter skin of northern Europeans versus the lighter skin of northeastern Asians. Even at the same latitude, skin color differs in a variety of ways, depending on how long the population has been living in that region and how much vitamin D is incorporated into the everyday diet.

In 2015, David Reich and his team at Harvard University published interesting research about differing skin tones. In a surprising discovery, it turns out the light skin of Europeans has been around for less than 5,000 years. This seems almost impossible; ancestral humans, upon leaving Africa, had to live in Europe during the Ice Age. Surely, with critically low levels of UV radiation at that time, they should have lost the melanin coloring that would have prevented adequate vitamin D synthesis. The mutation for white skin therefore should have originated at least several hundred thousand years ago, if not a million years ago or more, since this is when ancestral humans first started to live in Europe. And from there the mutation should have spread all over the world. Perhaps modern humans left Africa only recently and then migrated to Europe? Even in this more conservative scenario, the lighter-skin mutation should have appeared at least several tens of thousands of years ago in Neanderthals. Five thousand years is quite unexpectedly recent. What could have happened?

One hypothesis that is gaining traction suggests that this recent appearance has to do with agriculture and a sedentary lifestyle. Before agriculture, there was not as much need for us to synthesize vitamin D, even in areas with deficient UV radiation. Why? Because our everyday diet had a sufficient amount of vitamin D in the plants, marine resources, and meat we consumed. As we shifted to sedentary agriculture as our main source of subsistence, however, we increasingly relied on processed grains and starches, which were deficient in many nutrients, including vitamin D. With vitamin D through food no longer a viable solution, a mutation that caused the melanocytes (melanin-making cells) to become less active became advantageous. Less active melanocytes resulted in lighter skin, allowing for vitamin D synthesis even with a small amount of direct sunlight.

EXTRA

COMPLETE BLOCK IS NOT THE ANSWER

For a while now in the United States and Europe, tanning salons have been quite popular. As more research has been released showing that UV radiation is harmful to health, however, sunscreen use has gained popularity as sun lovers have started to worry about getting skin cancer. We've now become so good at using sunscreen, though, that its overuse is turning into a problem. Since 2000, the CDC (Centers for Disease Control and Prevention) has warned against the dangers of vitamin D deficiency brought on by the use of too much sunblock and not enough sun exposure. The CDC has also suggested that we start to add foods rich in vitamin D to our diets, such as milk and eggs. We know that too much of a good thing can be bad, but we need to be careful about where we draw the line.

CHAPTER 8

Granny Is an Artist

For most of human history, having a long life has been considered a blessing. Traditionally, not many people lived to be old. For example, until the first half of the twentieth century in Korea, a sixtieth birthday was a rare enough event that entire villages would turn out to celebrate. But only one generation later—barely twenty years—such large-scale sixtieth-birthday parties had become rare. Instead, many had moved the big celebration back to their seventieth birthday.

Nowadays, fewer and fewer people throw a big seventieth-birthday bash. People seem to be growing accustomed to their newfound longevity, waiting and looking forward to easily reaching a life span of a hundred years or more. But to tell the truth, the outlook is not as optimistic as you might think; long life also turns out to be a great source of anxiety. At the forefront of this anxiety is the emerging group of senior citizens who are increasingly frail. In our hearts, even though we are living longer, we know that we will not be as healthy in old age as we were when we were young. "When we turn ninety-nine, let's party like it's 1999" does not fill us with a sense of delightful hope; it sounds more like a desperate plea, because we know we'll inevitably be fragile at the end of such a long life. And the anxiety about growing old goes beyond our individual health; it has socioeconomic implications as well. This is why there's so much debate over Medicare and Social Security benefits.

With all the perceived drawbacks of aging, you might wonder why we ever evolved to grow older in the first place. Or are we missing another viewpoint on human longevity?

The rise of longevity

In part, longevity has increased because advances in modern medicine have helped reduce mortality rates. As recently as the beginning of the twentieth century, both fertility and mortality rates were relatively high (many people were born, and many died before reaching old age), making the average life span quite short. By the middle of the twentieth century, mortality had decreased as fewer people were dying, thanks to advances in modern medicine. Infant mortality, however, remained high. Many newborns died after only a few months. In traditional Korean culture, people refrained from congratulating pregnant women and even refrained from celebrating the birth. Instead, many families would wait until children were three months old to celebrate their birth in a ceremony called the *paegil*, which means "100 days." Such postponement of the celebration of birth can be found historically in many cultures and societies around the world, including Nigeria, Japan, Tonga, and Hawaii, where the first, third, or fifth birthday is a cause for a great celebration, rather than the day of birth.

The reason for delaying the celebration of birth was that until recently, many children were not expected to live past their first birthday. In fact, life goes through a series of mortality risks. Mortality increases around the time of weaning, regardless of when that happens, which can vary across different cultures. Increased reliance on food sources besides breast milk introduces children to a whole host of food-borne illnesses that can easily lead to diseases. Even after surviving this risky initial period of solid-food consumption, however, mortality increases again around sixteen to eighteen years of age, defined as the completion of somatic growth (growth of the body). After this time, young women face higher mortality due to risks associated with pregnancy and childbirth, while young men face higher mortality due to accidents. Middle-aged adults see yet another peak in mortality, this time due to diseases rather than accidents.

People go through numerous peaks and valleys of mortality risk throughout life. The increased average life span in contemporary times means that more people survive these risks of death; there are fewer accidents and wars, more diseases are treatable, fewer deaths result from complications of pregnancy and childbirth, and so on. With these obstacles to longevity minimized, but a high fertility rate persisting, the population had exploded by the end of the twentieth century. But then people in industrialized countries began to have fewer children; after all, the likelihood that the children they did have would reach adulthood had greatly increased. So now, for the first time in human history, most developed countries are experiencing extremely low fertility *and* extremely low mortality. As a result, we are presented with a new phenomenon: an increasing size of the elderly population. When we look at it this way, the longevity of humans appears to be a direct result of technological advances.

As we're about to see, however, our longevity did not increase just because of modern civilization. In fact, the number of elderly humans, specifically the number of grandparents, increased long before settled societies even started. Furthermore, we know that longevity is somewhat heritable: long-lived people tend to come from families of long-lived people. In fact, several specific genes, which had gone unnoticed until recently, are known to contribute to longevity. Could this mean that longevity actually has an evolutionary advantage?

Intuitively, longevity doesn't seem to be clearly advantageous. For a feature to have an "evolutionary advantage," it should contribute to leaving offspring behind; in other words, it should be advantageous for reproduction. Human females experience menopause, or reproductive senescence, around the age of fifty. Natural ovulatory cycles cease, and pregnancy with childbirth is no longer possible. Considering that evolutionary success is defined by reproductive success, there seems to be no evolutionary point to living long after menopause, since pregnancy and childbirth are no longer possible. Most females in the animal world remain reproductive until the end of their lives; the few that do reach menopause die shortly thereafter. In contrast, human females live an active, healthy life for at least ten to fifteen years after menopause.* Given that the life expectancy in 2017 for Korean women is ninety, women there may expect to live for forty more years after menopause. From the perspective of an average mammalian life span, this is a truly remarkable situation.

To solve the mystery of human longevity, anthropologist Kirsten Hawkes of the University of Utah proposed in 1989 the "grandmother hypothesis." According to this hypothesis, the elderly—specifically postmenopausal women—help with the survival of their genes not by giving birth themselves, but by helping to take care of the youngest members of their family. Through this mechanism, grandparents in particular are able to ensure the survival of their genes into later generations. Selection would favor grandmothers who have long, healthy lives past reproductive age.

When did longevity begin—with sapiens or erectus?

When did longevity first appear in human history? One recurring lesson in studying evolutionary history is that what seems to be a natural,

* Besides humans, a couple of whale species are known to have postmenopausal longevity.

obvious phenomenon is often the result of a long, meandering process. Such is the case with longevity. Proponents of the grandmother hypothesis argued that longevity first appeared 2 million years ago, with *Homo erectus*. The pillars of their argument were brain and body size: the bigger brains and larger bodies that set *Homo erectus* apart from previous hominins indicated a slow and lengthy aging process.

It was argued that the bigger brain and larger body were attained by delayed maturation. Maintaining a growth rate during development for a longer period of time will, of course, result in greater size. And if the growth rate for some parts of the body slows while the growth rate for other parts stays the same, those other parts will get even bigger. This slower and longer growth process was mirrored in old age, as a slow and lengthy aging process. If the grandmother hypothesis does indeed apply to *Homo erectus*, its aging process needed to be slow in order to maintain an active and energetic postreproductive phase.

As great as the grandmother hypothesis sounded in theory, it had a significant problem: it could not easily be tested using archaeological data or contemporary populations. One study showed that grandmothers were associated with decreased infant mortality; another did not. One computer simulation showed that grandmothering could not lead to longevity; another showed that it could. Some researchers have tried using osteological evidence (skeletal remains) from *Homo erectus* to estimate the age of death, but this task turns out to be very difficult with adult specimens.

It is relatively easy to estimate the age of individuals who did not complete their growth, because changes in bones and teeth are age-specific during the growth period. After puberty, however, there are no significant changes in bones or teeth that can be tied with specific ages. Moreover, there is so much individual variation in the aging process that assessing age is tricky with older individuals. For example, arthritis can be recognized in bones, but the same degree of osteoarthritis can be found in people in their thirties or in their fifties, so all one can say with accuracy is "thirty-plus years of age." These constraints present a serious challenge for testing the grandmother hypothesis through the fossil record.

Working with anthropologist Rachel Caspari of Central Michigan University, I decided to attack this problem from a new angle. If age assessment of adults can never be accurate enough to test the grandmother hypothesis directly, why not choose an indirect route? Instead of agonizing over accurate age estimation, we divided hominin fossil specimens into "young adults" and "old adults." We defined young adults as any specimens that completed the growth process and had a potential of reproduction, which is biologically marked by the appearance of the third molar (wisdom tooth). We defined old adults as those who had a potential to be a grandparent, and were at least twice as old as the youngest of the young adults.

We inferred "twice as old" from the degree of wear on the teeth, modifying the Miles method of age estimation developed in the 1960s. For instance, let's assume that the third molar appears at the age of eighteen on average, marking the initiation of young adulthood. A young adult who then began to have children would potentially be a grandparent around the age of thirty-six, when their oldest children would be turning eighteen and having children of their own. This is why we defined an "old adult" as someone with twice the wear (and therefore potentially twice the age) of a "young adult." If it was impossible to come up with an accurate age at death for adults, insisting on some specific age with an unknown error range would erode the credibility of the research; instead, we settled on a categorical approach that fit the characteristics of the data better.

We collected data from all hominin fossil remains that could be assessed using our "old" and "young" metrics—a total of 768 individuals. The data came from specimens that included australopithecines (belonging to the genus *Australopithecus*), *Homo erectus*, Neanderthals, and the European Upper Paleolithic *Homo sapiens*. We calculated the ratio of old adults and young adults (named the "OY ratio") for each group, to see whether it changed over different times and across different groups. A ratio of 1:1 meant there was an equal number of old adults and young adults, a ratio larger than 1:1 meant more old adults compared to young adults, and a ratio smaller than 1:1 meant more young adults.

As we expected, the OY ratio increased over time from the australopithecines on. The OY ratio for *Homo erectus* (starting 2 million years ago) was higher than that of the australopithecines (starting 4 million years ago), and the ratio in Neanderthals (starting 200,000 years ago) was higher than in either of the other two groups.

Our data revealed a surprise, though: the greatest degree of increase in the ratio, which can be called the "first appearance of longevity," did not coincide with the appearance of *Homo erectus*; it was associated with the European Upper Paleolithic *Homo sapiens*, starting about 30,000 years ago. Before this time, the OY ratio, even though it was increasing, never rose above 1:1. From the australopithecines until the Neanderthals, there were still more young adults than old adults. The *Homo sapiens* sample, however, exhibited an OY ratio of more than 2:1; there were twice as many old adults as young adults.

The increase was explosive. Considering that the European Upper Paleolithic period also marked an increase in burials, we wondered whether the burial cases could have biased the data, so we redid the analysis without data from burials. But the results were the same: the OY ratio in European Upper Paleolithic *Homo sapiens* was more than 2:1, representing a big increase from earlier time periods. The advent of grandparents, and human longevity, appears to have occurred during Upper Paleolithic culture, only 30,000 years ago, not with *Homo erectus* nearly 2 million years ago.

Longevity and the blossoming of art

Interestingly, the Upper Paleolithic culture, the culture of anatomically modern *Homo sapiens*, is considered revolutionarily different from previous cultures. Artistic and symbolic expressions, as seen in cave art and ornaments, began to flourish in this period. Was it a simple coincidence that this period also saw an explosive increase in longevity? Might there be a causal relationship between art and longevity?

Art and symbols are associated with more abstract thinking, and they also perform a practical function: the transmission of cultural information and meaning. The increased use of art and symbols during the Upper Paleolithic reflects the increasing importance of knowledge transmission. Intriguingly, longevity also contributes to increased knowledge transmission. Living long enough to see grandchildren meant that three generations could exist at the same time, so people could archive and share knowledge between generations for a longer period of time than earlier hominins could. If we assume twenty-five years for a single generation, and two generations share fifty years, then three generations can share seventy-five years' worth of cultural memory. Thus, longevity provided an actual mechanism for increased information production, sharing, and retention. As a consequence, it could very well have played a role in the birth of art and meaningful symbols.

Despite our increasing life span, this overlapping of just three generations does not seem to have changed much since the Upper Paleolithic. When the average life expectancy was barely sixty years, as was the case in the late 1970s for the world average, grandparents survived until their grandchildren grew up, leading to three generations living together at the same time. Since then, life expectancy has skyrocketed for some populations, as high as ninety for Korean women. Doesn't this mean there should be an increase in the number of great-grandparents who survive until their great-grandchildren reach adulthood? In other words, shouldn't there be an increasing number of cases of four generations coexisting?

Instead, there are fewer people at the age of sixty who have grandchildren, let alone great-grandchildren, than there were a couple of generations ago, even though many are now living past seventy years of age. What happened? It appears that young couples are delaying marriage and reproduction longer than ever before in our history. As a result, while a life expectancy of 100 years may be right in front of our eyes, we have preserved the Upper Paleolithic family structure of just three generations in coexistence (not four). Perhaps it is not that we're living *longer* than in previous times, but rather that we're just living more slowly. The age of a "Slow Life" is indeed upon us.

EXTRA
HOW LONG CAN YOU LIVE?

Although we are about to enter the age of centenarians (100-year-olds), it doesn't mean that our average *maximum* life span is longer. We call that our "absolute life span"—in other words, our possible life span in the absence of accidents or diseases, determined only by our aging process. No matter how much modern medicine advances, our absolute life span has an upper threshold.

How long is an average human's absolute life span? Although we do not know for sure, we hypothesize that it should be approximately the same as the median of the longest recorded life spans. Currently, the oldest person on record is Jeanne Calment (1875–1997), who died at the age of 122. Furthermore, the top 100 individuals holding the longest life spans recorded are all concentrated at about 114–119 years of age. Among these people, only 8 have a chance at setting a new record of longevity. This seems to be indirect evidence that absolute life span does not increase, no matter how many advances are made in modern medicine. The "era of centenarians" does not refer to increased maximum life span, but rather to increased survivorship of adults into old age.

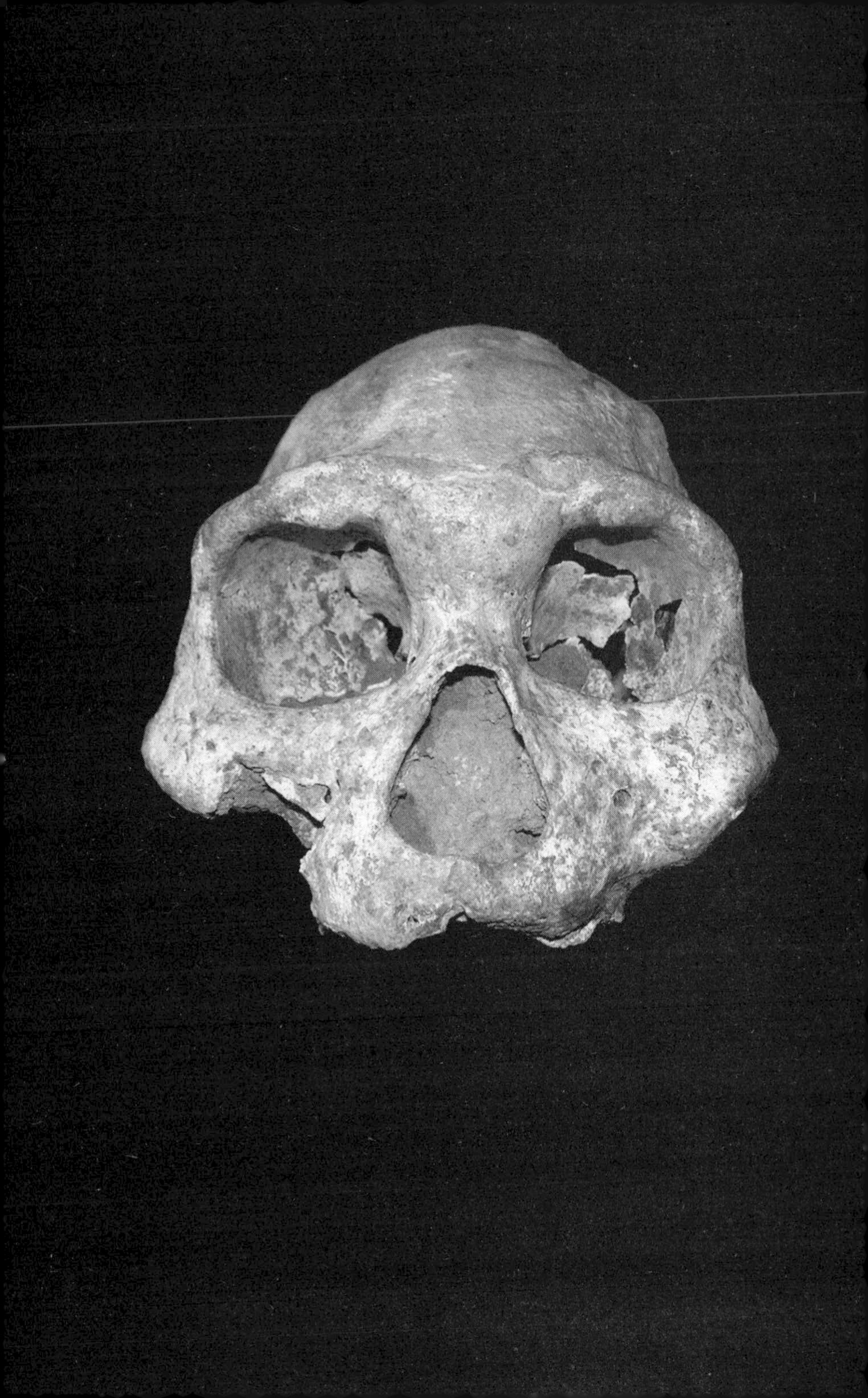

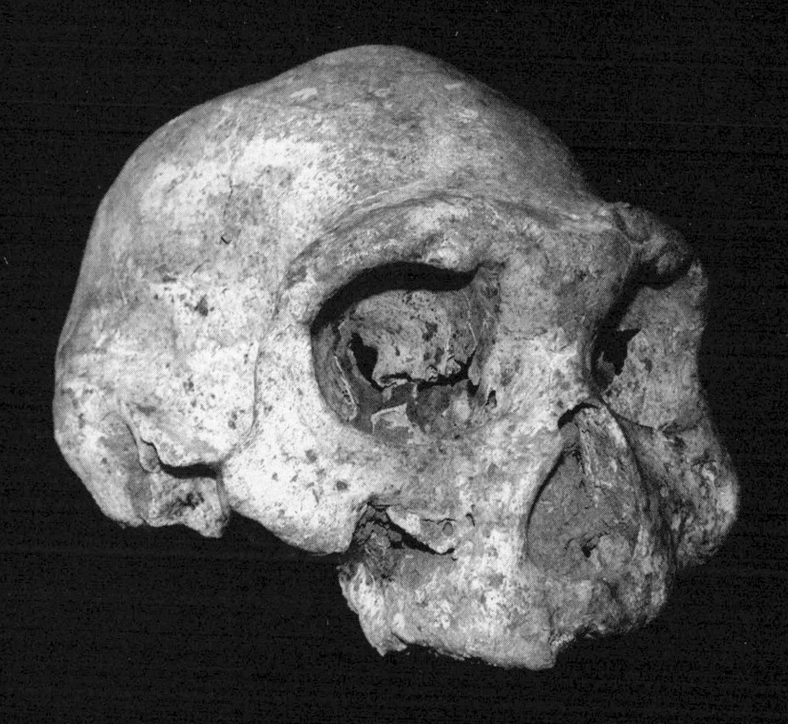

Two views of fossil skull of *Homo georgicus* discovered in Dmanisi, Georgia. (© Milford Wolpoff)

CHAPTER 9

Did Farming Bring Prosperity?

In many parts of the world, agriculture has long meant prosperity. In Korea, agriculture and farmers were traditionally considered the "foundation of the world" and were held in high regard. After foraging for natural resources for several million years, humans started about 10,000–11,000 years ago to systematically grow plants and domesticate animals on a large scale to produce food for themselves. In retrospect, this seems to have been the dawn of prosperity for humanity. If "primitive" societies had to wander around all day to get the minimum amount of food necessary to keep living until tomorrow, farmers, by contrast, seemed to enjoy a rich lifestyle, working just a little and harvesting a lot. Without needing to do so much work, humans began to have leisure time. Without having to move around looking for food, humans began to live together in one place. With agriculture, we could expect to look forward to enjoying a long life with a disease-free, healthy body, living with family and friends in the brilliant civilizations that began to blossom. Or so we thought.

Anthropological and archaeological research spanning the last fifty years has brought to question what we believed about the impact of agriculture on the human body and society.

Farming is unhealthy

To study life before agriculture, anthropologists lived together for several years with modern-day "primitive" societies that do not farm, conducting what is called ethnographic research. The most prominent example is the Harvard Kalahari Project, involving several different teams living in Africa between 1950 and 1979. Admittedly, the exposure sometimes resulted in an unfortunately simplistic, misleading, and racist picture of the Kalahari people, better known as the San or !Kung. For example, the movie *The Gods Must Be Crazy* (1980) is a comedy premised on the discovery of a Coca-Cola bottle by these "bushmen."

Continuous accumulation of solid research, however, led to an in-depth understanding of the hunter-gatherer cultures of the !Kung and brought to light surprising facts about their life, which was rich and decidedly not "primitive" or savage. Although they could not afford to lead a completely work-free life, the !Kung had substantial leisure time. They were not roaming around all day, half-starved, nor were they subject to widespread famine or infectious diseases. The absence of agriculture did not make life brutal and difficult.

Does this mean, then, that an already rich hunter-gatherer lifestyle just got richer with agriculture? No, not really. Instead, studies of human skeletal remains revealed that once-healthy people became vulnerable to many diseases and malnutrition only *after* they adopted agriculture.

Human skeletons and teeth tell us a lot about a person's childhood, life, and ailments. Without sufficient nutrition in the early years, normal growth patterns are interrupted, leaving a signature on the bones and teeth. Enamel hypoplasia is a well-known example of this type of signature. The condition is caused by deformity in the enamel of permanent teeth from malnutrition during childhood. Since permanent teeth are made once in a person's lifetime, the signs of nutritional ailments that occurred during childhood remain permanently etched in those teeth. Anthropologists discovered that the frequency of enamel hypoplasia increased conspicuously when a population adopted agriculture, suggesting that adopting agriculture led to severe malnutrition in a population.

Similarly, measurements of limb bones (arm bones or leg bones) show that those of agriculturally based populations were shorter. With the advent of agriculture, people became shorter than their predecessors, because of bouts of famine and severe malnutrition.

These findings show how erroneous it is to assume that agriculture brought a rich and prosperous lifestyle. In fact, the malnutrition that has accompanied agricultural lifestyles in the past can even be seen today. Have you seen pictures of malnourished young children with swollen stomachs? This symptom is a sign of a disease called kwashiorkor. Contrary to what you might expect, this disease is not caused by a deficiency in caloric intake. Instead, the cause is insufficient protein intake, even when overall caloric intake is sufficient. In other words, you can get this condition if you eat only nutritionally deficient foods (starches or processed, empty carbs) every day. Left untreated, this disease can be fatal. Ironically, overall malnutrition, where total intake, not just protein intake, is deficient, is less dangerous to the human body.

Why are there health risks when we produce food directly from the land? In some ways, agriculture is similar to an investment plan that focuses on stocks from just one or two companies, rather than a diversified portfolio. If the weather and soil is good, the return on farming investment is large, and everyone can have a feast. If the weather is unfavorable, however, the harvest for the whole year can be ruined, and everyone will be starving the following year. In contrast, with the foraging economy before agriculture, we acquired food from a broad territory and a diverse range of sources. Even when one particular food item had run out, there were many alternatives. To be sure, feasts were rare, but so were famines.

Humans clearly did not eat better after adopting agriculture. To make matters worse, diseases also increased and ran rampant among early agricultural societies. Take dental cavities and gum diseases, for example. In many agricultural societies, water is added to the staple grain, which is then cooked, much like rice or pasta, to become a soft cereal. A diet like that has a higher risk for dental cavities than does a diet with hard, abrasive foodstuffs, because the sticky starch is more likely to remain on the teeth and provide nutrition to the bacteria that cause dental caries. Nowadays, with advances in modern dentistry, we have no real concept of how serious dental cavities could be. In early societies before modern dentistry and dental hygiene practices, dental disease was devastatingly painful. If the infection spread, teeth would have been lost, and infection in the gums could have spread to the whole body and become fatal. And did I mention it was devastatingly painful?

Furthermore, a sedentary life, one of the essential aspects of agriculture, leads to vulnerability to infectious diseases. As humans became bound to the land, they could not easily leave a place, even if a disease with a high mortality rate was spreading. Furthermore, living close to one another meant that once one person contracted an infectious disease, it was likely only a matter of time before the whole village would succumb to the illness and it would spread to nearby villages. When humans were nomadic, infectious diseases would be quickly left behind when we moved.

Community life had other implications for disease beyond simply rapid transmission. Pathogens suddenly lived in a rich environment where new hosts were supplied almost infinitely, house after house. The result was an evolutionary change in the pathogens themselves: they could afford to be more virulent and deadly. Before human societies became sedentary, virulence was not beneficial for pathogens. To survive in a highly mobile human population, pathogens had to live with their hosts for a long time, to ensure that they wouldn't be trapped in a host body that suddenly died. Thus, human-borne pathogens were evolutionarily selected for weak virulence, and infected humans survived for a relatively long time. But in this new, settled-society environment, when a host died, a new one nearby could immediately take its place, over and over again. Pathogens could afford to be strongly virulent; they could afford to kill their hosts.

Now consider the addition of domesticated animals, which were likewise subject to this sedentary lifestyle. Diseases of animal origin began to evolve trans-species strains, also called "zoonoses," that could jump to human hosts. Humans soon were succumbing to attacks by a new breed of formidable diseases.

Population explosion: boon or bane?

Judging by the narrative up until now, we might expect a substantial decrease in population to have accompanied the onset of agriculture. But a strange thing happened: human populations increased, almost explosively.

Even more surprisingly, this increase was *not* due to a decrease in mortality. Mortality actually increased with the rise of agriculture because of the increase in mortality-contributing factors, such as infectious diseases. The population explosion was due not to a decrease in deaths, but rather to a large increase in births. In other words, fertility far surpassed mortality. And we have agriculture to thank for that.

While humans were foraging and mobile, the interbirth interval (spacing between siblings) was approximately four to five years. It would be more manageable for the mother to have her next baby after the older one could walk about and eat on its own. It is extremely difficult to raise two babies who need to be carried and cared for all the time.

You might be wondering how humans could control interbirth

intervals without the benefit of modern medicine and contraception. Interbirth intervals in nature often rely on the weaning period. Females do not ovulate during intensive breastfeeding because the hormone in charge of lactation suppresses ovulation—a phenomenon known as lactational amenorrhea. Once a baby is weaned and the lactational hormone decreases because breast milk is no longer needed, the female body resumes ovulation and menstrual cycling.

Nonagricultural ethnographic populations often breastfeed for three to four years, after which another pregnancy occurs. This pattern results in an interbirth interval of four to five years. With a new farm-based diet relying on grains and starches, however, everything changed. Babies could be given new weaning foods—cereal and gruel—instead of their mother's breast milk. As a result, the baby could be weaned and leave the mother's care much earlier, and the mother's body could then prepare to have another baby. Now a woman could raise babies with an interbirth interval of just two years.

As fertility increased substantially, the population grew rapidly. In evolutionary biology, population increase in a species is a mark of its successful biological adaptation. Clearly, this success was thanks to a sedentary lifestyle, agriculture, and cooking.

Does our population explosion show that agriculture was ultimately an unmitigated successful evolutionary adaptation? Not quite. Rapidly increasing populations led to yet another tragedy. More farmland became necessary to feed larger populations. As a result, humans started to engage in small- and then large-scale wars to take control of more land. Wars increased mortality. As mortality increased, there were fewer and fewer people to send off to war *and* also work the land. The need for more children continued to rise. Now, women's lives consisted of having to have another baby while an older baby was still very young, all while working hard to plow the land.

As population and productivity increased, surplus was generated. Divisions arose in which a segment of people controlled the distribution of surplus, and the distribution power was inherited through family lines. Agricultural society became stratified with a highly complex and detailed class system. Cities, nation-states, and civilization ensued. But we still have this question: With agriculture, did we step closer to prosperity or farther away? The late George Armelagos, an anthropologist at Emory University, is famous for posing this question. And the quote often attributed to him—"Agriculture is the biggest mistake in human history"—probably has some merit.

Farming's genetic gift

Contrary to what we commonly understand, agriculture was not solely a blessing to humanity. But it was not all bad news either. Genetics, in particular, has revealed a new reason for us to appreciate the hitherto hidden contribution of agriculture: genetic diversity. Thanks to agriculture, population expansion triggered many more opportunities for genetic mutations to occur within the population, increasing our genetic diversity. Along with human population, the raw material for evolution also explosively increased.

Usually, the word "mutation" has a negative connotation, but mutations are an integral part of evolution. The modern definition of evolutionary success rests on the idea of increased reproductive fitness—of making as many genetic copies as possible, in the form of offspring. Mutations in genes, which often occur during the copying process, may lead to different traits. If these new traits give the organism an evolutionary advantage, they are passed on, along with the rest of the organism's genetic material, at a higher rate than that of organisms without the new traits. The species will have a higher frequency of the advantageous new traits through this process, called "natural selection."

As such, mutation brings variation in traits and becomes the raw material for evolution; more opportunities for mutation means more opportunities for evolutionary success. Mutations occur randomly. Let's say a mutation occurs with a rate of one in a 1,000 genes. Then there will be 10 mutations in every 10,000 genes, 1,000 in every million, and so on. As the human population exploded with agriculture, mutations also increased, leading to an explosion in genetic diversity. We can trace most of the great diversity that we see in humans today back to agriculture.

The role of agriculture in increasing genetic diversity is a significant event in human history, not only for its contribution to our evolutionary success, but also because it is a case in which "civilization" directly influenced human evolution. For a long time we thought that evolution stopped when culture and civilization developed. Agriculture shows that culture and civilization can have direct, and dynamic, effects on human evolution via population explosion (see Chapter 22).

Nowadays, as discussed in the previous chapter, humanity is confronting yet another cultural phenomenon for the first time in history: the increased longevity of the elderly. If culture directly influences human evolution, this older population will surely lead us in a new evolutionary direction. How will humanity respond to this new phenomenon?

CHAPTER 10

Peking Man and the Yakuza

In the fall of 2009, I visited Beijing, China. I was invited to a conference commemorating the eightieth anniversary of the discovery of Peking Man. When the conference was over, I had the opportunity to visit the famous site of Peking Man's discovery: Zhoukoudian, a cave southwest of Beijing. As I explored the cave, I was awestruck. Not just because it was a site of utmost importance in the history of paleoanthropology, but also because it reminded me of a strange email I had received ten years before: an invitation to go on a "ride-along" with the notorious crime organization of Japan, the yakuza.

At the time, I was a postdoctoral researcher at a genetics institution in Hayama, Japan, south of Tokyo. The stranger who sent the email introduced himself as a journalist who had followed the yakuza all his life. He went on to explain that the following week there was going to be an initiation rite for the yakuza, and he would like me to accompany him to the event. At first, I was baffled by the idea that a paleoanthropologist was needed at a yakuza initiation ceremony. As I read further, however, the situation became clear: the journalist had received a tip that the original fossil of Peking Man was supposedly going to appear at this secret initiation rite, and he wanted an expert with him to assess whether it was the real, original, Peking Man fossil. My curiosity was piqued. If the claim was legitimate, it would be a truly rare moment in the history of paleoanthropology. It would also be a chance to solve a decades-old cold case: the famous Peking Man fossil had been missing since World War II.

The Peking Man mystery

Peking Man was discovered in Zhoukoudian, China, in the 1920s. The find consisted of numerous fossils, including several partial crania, mandibles, numerous teeth, and some postcranial (below-the-neck) skeletal bones. They are historically significant, along with the Java Man fossils discovered in Indonesia at the end of the nineteenth century, as the earliest evidence of hominin presence in East Asia. The excavations started with a single molar and continued until 1937, when the Japanese invasion of China interrupted the work. In 1941 the fossils were taken to the Qinhuangdao harbor, in Bohai Bay east of Beijing, in preparation for transport to the United States, where it was hoped they could be safely stored until the end of the war. That was the last time they were seen; they vanished from the harbor without a trace.

Ever since, paleoanthropologists have been searching for the original Peking Man fossils. Several hypotheses have been proposed as to their location. Some have argued that the CIA has them; others have claimed China has them. There was great excitement in China in 2012 when a newspaper published an account from a witness who claimed to have long ago seen the box that contained the original fossils. The witness contacted a scientist, who subsequently concluded that the box containing Peking Man had most likely been destroyed in the bombings during World War II. Even if it hadn't been destroyed, it would now be buried deep under the new roads that had been laid down as the region was developed into a port city.

Since the whole story of the mysterious disappearance and then reappearance of the box seemed extremely dubious, it was highly unlikely that the city would demolish a whole port in search of the improbable box buried underneath it. This story gathered such attention, however, that the project was eventually undertaken by the National Geographic Society, and the Chinese government supported the investigation. As of 2017, nothing seems to have come of this effort, since no newsworthy discovery has been reported.

Another of the many theories about the Peking Man fossils claimed that the yakuza had acquired the fossils through illegal means—hence the need for a paleoanthropologist at the upcoming yakuza ceremony. I trembled with excitement at the thought of being able to contribute to the rediscovery of Peking Man's remains. I sent an email to my adviser in the United States, asking for his advice. He immediately wrote back "NO!!" He was vehemently opposed to the idea, pointing out the extreme physical danger I would be putting myself in if I accepted the invitation. So, after much internal conflict, I declined the journalist's invitation.

Perhaps it was just as well in the end; if I lacked the courage to go against the recommendation of my adviser on the other side of the Pacific, how on earth would I be able to confront the yakuza face-to-face?

Homo erectus: *strong fire wielders or weak wanderers?*

The Peking Man fossils have never been rediscovered. But their disappearance has not stopped scientific inquiry, thanks to the German anatomist Franz Weidenreich, who made detailed casts of the original

Peking Man. His molds are of such high quality that they could almost substitute for the originals. Many scholars have continued to conduct research about the life of Peking Man.

The Beijing region is a cold and difficult place to live, and 500,000 years ago, during the Ice Age, it was much worse. To survive in a place like that, *Homo erectus* (the species designation of Peking Man) had to adapt culturally by living in caves, making fire, and wearing warm furs. Discovered in the Zhoukoudian cave along with Peking Man was a circular layer of ashes, an artifact of a fire having been made. Also discovered in the cave were various animal bones and stone tools. You can almost picture the original scene: a deep valley during a bone-chilling blizzard with a group of people huddled around a warm fire in a warm cave, eating roasted meat and sharing stories into the long night. *Homo erectus* from the Zhoukoudian cave was thought to look and act like modern humans in this way, and therefore this image was firmly set in people's minds.

Recently, the life of *Homo erectus* has been undergoing reexamination, however, because *Homo erectus* fossils are continually being discovered in China and the rest of the world. In particular, the idea of "controlling fire" has come under scrutiny. There is no doubt that *Homo erectus* used fire. What is unclear is whether individuals of the species controlled and maintained fire, making it whenever needed, or whether they were simply opportunistic *users* of fire, cooking and warming themselves whenever there happened to be a fire nearby. The distinction is key to placing *Homo erectus* in the right spot along the continuum toward modern humans. If members of this species could make fire whenever they wanted, if they controlled fire, they would be that much closer to humanity.

Susan Antón, an anthropologist at New York University, makes a particularly interesting observation on this matter. She argues that Peking Man lived in the Zhoukoudian cave not during the colder (stadial) period, but rather the warmer (interstadial) period of the glacial Ice Age. Furthermore, she argues that Peking Man is not representative of *Homo erectus* fossils in general, but is an exceptionally unusual sample of the hominin lineage. In other words, Peking Man is a somewhat alien offshoot of *Homo erectus* that accidentally ended up in Zhoukoudian. It is indeed ironic to think that Peking Man, the hallmark example of Chinese *Homo erectus* for several decades, may not be representative of the populations living in the region at the time. The representative *Homo erectus* may not be so representative after all.

So, who is this *Homo erectus*? Modern humans (*Homo sapiens*) living on the Asian continent today can be divided into two groups: the inland population of the north, and the coastline population of the south. The same might have been true of *Homo erectus*. Which of these two populations did Peking Man belong to? Were these hominins people of the north who endured the long glacial winter with furs and fire, or people of the south who wound up in the cold inland of northeastern Asia while moving during the warmer interstadial period? Mainstream opinion is with the northern hypothesis, but Antón's research cautiously supports the southern hypothesis.

Of course, nothing is certain. Neither position is supported by rock-solid evidence. This mystery seems to be yet another example of how no hypothesis in science can be an absolute truth. While we may know quite a lot about Peking Man, much of the hominin's mysterious evolutionary history is just as mysterious as its disappearance from Beijing's harbor.

The mystery continues

So, whatever happened to the original Peking Man fossils that the Japanese journalist was going after? As I learned more about how formidable the yakuza are, I grew very scared. I deleted all the emails from my exchange with the journalist. Since no news reached me of a discovery of the original Peking Man fossils in Japan, I guess the journalist's ride-along did not go well, or the information was wrong. Either way, the mystery continues.

Ten years later, I was thinking of this time when I was standing in the Zhoukoudian cave, feeling nostalgic. Where on earth are the original Peking Man fossils? Are they really in the hands of the yakuza? Are they hidden in the United States? Perhaps Peking Man witnessed the end of the war and the rapid economic development of China from its grave under the harbor road. Wherever the fossils are, I wonder if they miss home and the memory of roaming the continent in East Asia, 500,000 years ago. After all, they are likely the group of *Homo erectus* that traveled the farthest of their species.

EXTRA

"FACELESS" PEKING MAN AND CANNIBALISM

Some of the stories told about Peking Man are quite scary. For example, some say that members of the Peking Man group were always starving, and sometimes killed others and ate them. This conjecture lacks any evidence to support it. There is, however, one curious fact about the Peking Man fossils: very few facial remains have ever been found. In fact, many hominin fossils from Asia, including Peking Man, have only the skullcaps remaining, while their faces are completely gone.

Facial bones are unlikely to be fossilized, because they are small, thin, and fragile. Nonetheless, even among hominin fossils, Peking Man fossils have an unusually small number of facial bones. For example, hominin fossils from Europe and Africa are found with faces more often than

Peking Man fossils are. Why do hominin fossils from Asia lack faces? Is cannibalism the explanation? Would eating others have been the only way to survive in the blizzard-prone, icy valleys of northeastern Asia?

As discussed in Chapter 1, cannibalism cannot be a sustainable dietary choice, so there must be another explanation. Some people argue that Peking Man individuals were extremely violent, but not necessarily cannibalistic. This conclusion is based on the fact that Peking Man fossils have thick skull bones, which perhaps were an adaptation to a violence-prone lifestyle. But as more fossils from a similar time frame were found in other parts of the world, it became clear that thick skull bones were not specific to Asia, but common all over the world. Now, neither the argument that Peking Man was inherently violent nor the speculation that he was cannibalistic has much support in the field.

CHAPTER 11

Asia Challenges Africa's Stronghold on the Birthplace of Humanity

As of 2015, the tallest building in the world is Burj Khalifa in Dubai. It is more than 2,700 feet high, with 163 floors. This massive feat of engineering was challenged by China. The People's Republic declared it would erect a building that would stand at 2,750 feet, with 220 floors. Even more surprising, China proposed to build it in ninety days. Both in height and speed, this building promised to set a world record.*

In addition to having the tallest building, China seeks to claim another world record: the origin of the first hominins. One might scoff at the idea, knowing anything about human evolution.

As mentioned several times throughout this book, the first hominins appeared in Africa at least 4–5 million years ago. If we include even earlier species, such as *Sahelanthropus tchadensis* and *Orrorin tugenensis*, the time of origin may be as early as 6–7 million years ago. Regardless, the place of origin is indisputably Africa (see Chapter 3). Many in the field support the theory that anatomically

* Because of mounting safety issues, however, construction was stopped in 2013.

modern humans also originated from Africa. *Homo erectus,* which lies somewhere between the first hominins and modern humans, most likely originated in Africa too.

The mainstream position is that the most important events in human evolution happened in Africa. Only recently, however, was the field able to reach such a consensus. Many countries, including China, argued that they were *the* place of humanity's origin. As recently as 1975, Chinese research teams were publishing reports of the discovery of an *Australopithecus* fossil specimen found in China, but such publications have not received much attention outside of Asia.

Although China's claim to be the birthplace of humanity is probably without basis, the question is far from settled. Human evolution has never been straightforward; *Homo erectus,* the ancestor of modern humans, might, in fact, have originated in Asia.

Java Man

As the theory of evolution became widespread toward the end of the nineteenth century, humans started to come to grips with a new idea: perhaps we did not appear on Earth in the shape and form we have today. In other words, perhaps we did not first appear as perfectly formed humans. People began to accept the notion that there were ancestral humans, descended from apes, that looked slightly "less human." These hypothesized ancestors would have looked like an intermediate step between apes and humans, but they were thought to have only the best traits of both.

Eugène Dubois, a Dutch paleoanthropologist and anatomist, believed that such an intermediary form not only existed, but would be found in Asia. In fact, Dubois was so convinced of this idea that, starting in 1887, he used his own personal funds to support excavations in the tropical forests of Indonesia. He thought that the earliest ancestral humans must have lived in the same ecozone as modern apes, so he predicted that fossils would be discovered in the woodlands where apes live now.

Then, in 1891, his excavations uncovered a hominin fossil on the island of Java, Indonesia. It was a revolutionary finding. So many paleoanthropologists spend their lives excavating, yet never succeed in discovering hominin fossils. The fact that Dubois discovered a hominin fossil at his first excavation site was nothing short of miraculous. Such good fortune would be matched by few in the history of paleoanthropological research.

The fossil that Dubois discovered was nicknamed "Java Man" after its place of origin. Java Man consisted of a skull, a tooth, and a femur. The skull was small and flat, while the femur looked like that of a modern human. So, Java Man had a brain smaller than that of modern humans but could walk on two legs. Dubois named the fossil species *Pithecanthropus erectus,* meaning "upright-walking ape-man." This species would later become reclassified as *Homo erectus.**

After such a discovery, we would expect Dubois to have been acknowledged by the field of paleoanthropology as the one to discover the first direct ancestor of humanity. Things were a little different in the late nineteenth century, however. Many people felt uncomfortable with the idea that humans, the smartest beings on Earth, could have an ancestor with such a small brain. The consensus was that such an unintelligent hominin could not possibly be an ancestral human, no matter how well it walked upright. Dubois's discovery was mostly ignored and forgotten by the field and by society, and he lived out the rest of his life in despair.

* Fossils do not come out of the ground with a label of genus and species names. The names given by scholars may be continually reexamined and reclassified. This is why species names change.

A race for the place of the first direct ancestor

With Dubois's Java Man discounted, the title of "first direct ancestor" to humans remained unclaimed until the early twentieth century, when groundbreaking discoveries were made simultaneously in Europe, Africa, and Asia.

In Europe, a specimen was discovered in Piltdown, near London, by Charles Dawson. Piltdown Man was welcomed with excitement, because it looked exactly like what paleoanthropologists at the time expected ancestral humans to look like, and it was initially given the species name *Eoanthropus dawsoni*, which means "Dawson's dawn-man." The fossil find had a large, round skull and large teeth, hinting at an intimidating presence equipped with excellent intelligence, ferocious teeth, and a robust body. That such a wonderful ancestor was found near the capital of England must have been comforting to the people of Great Britain. Yet a persistent rumor popped up immediately that Piltdown Man might be a hoax. Then finally, in 1953, it was indeed proved to be a fake, created by putting a human and ape skull together.

The second discovery was a small fossil skeleton named "Taung Child," found in South Africa in the 1920s. It was *Australopithecus africanus*, a new fossil species discovered by Raymond Dart, an Australian paleoanthropologist and anatomist. Currently, this species is considered one of the possible direct ancestors of early hominins. Nevertheless, this discovery was also ignored and dismissed at the time, much as Java Man had been.

The reason may have been that Taung Child was from Africa, a land that Europeans considered primitive and barbaric. The Western scientific community could hardly accept the idea that a species as brilliant as humans had ancestors from the African continent. What was even more damning was the fact that *Australopithecus* had a brain the size of an adult chimpanzee's brain. In addition, the specimen lacked any evidence of stone tool use and had modest, unimpressive teeth.

None of its traits appeared to show continuity with the excellence that Europeans believed themselves to possess.

Finally, paleoanthropologists considered a third candidate: Peking Man, discovered in China in the 1920s, in the Zhoukoudian cave, near Beijing. Peking Man was first classified as *Sinanthropus pekinensis,* which means "Chinese man from Beijing." Then it was reclassified as *Homo erectus,* the same species as Java Man. Although the complete disappearance of the original fossils remains a mystery (see Chapter 10), excavations of the cave continued to yield rich data. The excellent fossil casts made by Peking Man's original researcher, Franz Weidenreich, have been used for additional research as well.

Peking Man had a bigger brain, twice as big as that of *Australopithecus* (but approximately two-thirds the size of a modern human brain). The big brain of Peking Man led us to expect humanlike behavior and life from the species. That expectation was reinforced by the animal bones, stone tools, and ash layers found in the Zhoukoudian cave as well. It's estimated that Peking Man lived about 500,000 years ago. That such humanlike ancestors lived as far back as half a million years ago in China has been an immense source of pride for the Chinese.

Peking Man soon became representative of the *Homo erectus* morphological type. Some Chinese scholars argued that the earliest direct ancestors of humans originated in China. There was a problem, however, with the Chinese-origin argument for the earliest ancestral humans: earlier hominins much older than *Homo erectus,* such as the australopithecines, were found only in Africa. It is difficult to imagine how the *Australopithecus* genus in Africa was connected to *Homo erectus* far away in Asia.

This debate seemed to be settled when East Africa yielded *Homo erectus* fossils in the 1970s. These African hominins had a brain similar in size to that of Peking Man, with a body as big as that of modern humans, and were dated to 1.5–2 million years ago. The find presented a new scenario for the recent origin of humans. *Homo erectus* had originated in Africa first, equipped with a large brain, tall body, and excellent hunting tools, and then had slowly spread all over Eurasia. *Homo erectus* specimens found in Europe and Asia, such as Peking Man and Java Man, were now considered part of this migration that originated in Africa. Moreover, this revised scenario seemed to fit perfectly with the chronology and geographic distribution of hominin fossils across the globe.

Are we from Africa or Asia?

A big surprise was in store, however. In the 1990s a team of scientists reported that the Java Man remains are as old as 1.8 million years. The new date for the Java Man fossils was controversial at the time, and it remains so today. It means that the appearance of *Homo erectus* in Asia more or less overlapped in time with the existence of *Homo erectus* in Africa, suggesting that either Java Man might not represent a direct descendant of the African *Homo erectus*, or that an earlier *Homo erectus* specimen would need to be found in Africa to further confirm the African-origin scenario.

Then, a new discovery precipitated a stronger and more certain reversal in the field. In Dmanisi, Georgia, the country northeast of Turkey, a set of strange fossils was discovered in 1991. These fossils did not have big brains or big bodies. And the stone tools that were found with them were not so sophisticated. Paleoanthropologists scratched their heads. The African-origin hypothesis could not explain these modest-looking fossils being discovered outside of Africa.

Furthermore, the dates did not fit. The Dmanisi remains were as old as the first *Homo erectus* found in Africa, 1.8 million years. What did this mean? One could posit a new scenario: Perhaps this ancestral hominin species, with a small brain, small body, and crude stone tools, lived in Africa before *Homo erectus*, and left Africa in a separate migration. Perhaps it traveled through the Caucasus and ultimately reached Java, Indonesia. And then this line went extinct, except for a group in Asia that evolved into *Homo erectus*. *Homo erectus* then spread across Asia and all over the Old World, including migrating back into Africa. By this line of reasoning, the African *Homo erectus* would be a descendant of the Asian *Homo erectus*. Robin Dennell of Sheffield University, England, is a well-known proponent of this Asian-origin hypothesis.

It is important to keep in mind that this is only a hypothesis, not a proven theory of human origin. The point is that, with the discovery of the Dmanisi fossils, the hypothesis of a non-African origin for *Homo erectus* deserves more attention than our passing apathy. The Asian-origin hypothesis can no longer be easily dismissed. We are all paying attention to how valid this hypothesis is as our knowledge about human origins continues to grow.

EXTRA

PILTDOWN MAN, THE INFAMOUS HOAX IN PALEOANTHROPOLOGY

As mentioned in this chapter, Piltdown Man is one of the best-known hoaxes in recent scientific memory. In 1912, pieces of a skull, a mandible (jawbone) with an attached molar, and a canine were discovered in Piltdown, East Sussex, England. As Charles Dawson, a fossil hunter, announced his great find, people welcomed it as a discovery of the "missing link" between apes and humans. Yet many questions were left unanswered.

Several morphological traits did not make sense, given the evolutionary trajectory of apes and humans known at that point. The brain size was reconstructed to be between those of apes and humans (two-thirds the size of the modern human brain), but the braincase had more traits of a modern human than of ancestral human fossils. Because Piltdown Man was compatible with the idea of ancestral humans with big brains and ferocious teeth, however, for another forty years it was regarded as the first ancestral human. Finally, a thorough scientific investigation led to the conclusion in 1953 that it was a hoax.

The method that showed Piltdown to be a hoax was fluorine dating, which consequently became a prominent method. Fluorine dating is a relative dating method that assesses the difference in age between two or more objects, rather than giving a numerical estimate of an object's age (as an absolute dating method like radiocarbon dating does, for example). When a living organism dies and is buried, fluorine from the surrounding soil starts to accumulate in the body. The older a bone is, the more fluorine it has. Examination of Piltdown Man showed that the amount of fluorine in the skull was different from that in the mandible. This difference meant that the owner of the skull and the owner of the mandible did not die at the same time; in other words, the two bones did not belong to the same individual. As it was later revealed, Piltdown Man had been constructed from a medieval human skull, the mandible of an orangutan from 500 years ago, and a chimpanzee canine from an unknown date.

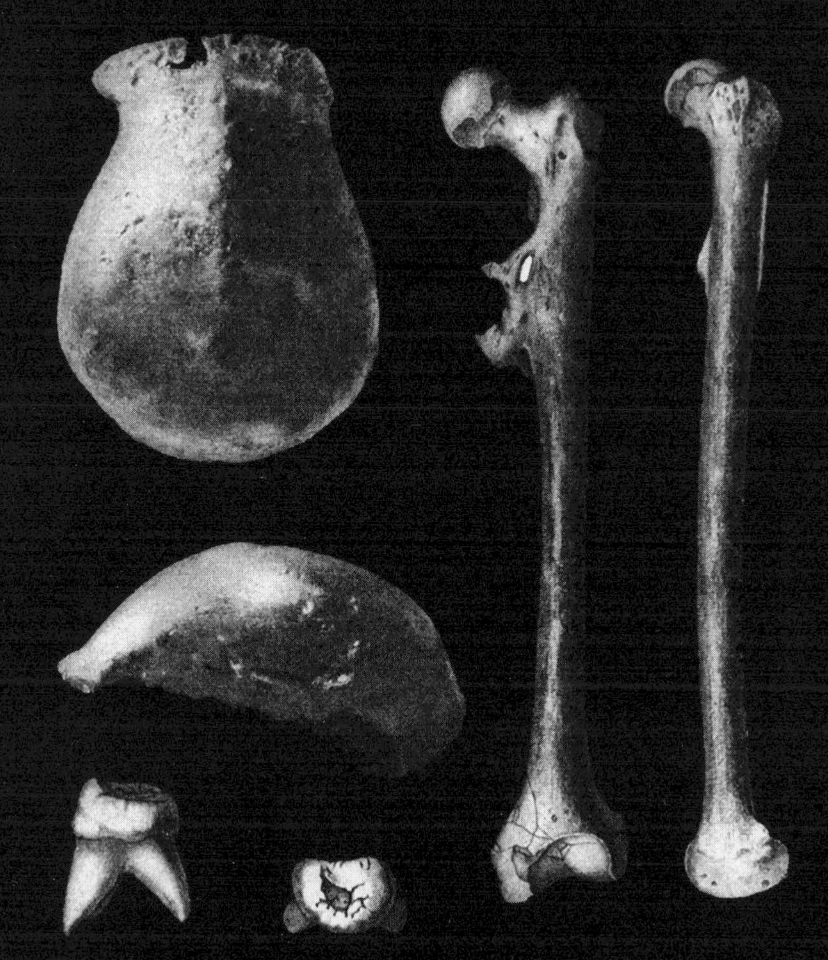

Drawings of *Homo erectus*, "Java Man," discovered in Java, Indonesia.

CHAPTER 12

Cooperation Connects You and Me

In 2012 there was a mass shooting at Sandy Hook Elementary School in Newtown, Connecticut. Twenty first-graders and six teachers and administrators were killed. It was a horrible tragedy. As the days passed, stories emerged of heroic acts. The principal had thrown herself against the perpetrator to protect students. A teacher had led the perpetrator in another direction after hiding children in a closet. Another teacher had even charged toward the perpetrator to prevent him from harming others. All of these heroic acts were carried out by female teachers who had no combat training. What spurred them to act in such a heroic manner? Considering that all the heroes were women, some might say it was simply maternal instinct in action.

Here is another example. Major Kang Jaegu, a squadron leader of the Korean army, threw himself on a grenade that his subordinate had activated by mistake, to save his squadron members during training for the Vietnam War. In this case, some might say heroism is to be expected, since signing up for the military service to go to a war zone means being ready to risk one's life for others. But the Sewol ship tragedy that shocked Koreans in 2014 is harder to explain. A Korean ferry carrying 476 people sank, drowning 304. On board were 325 students from the junior class of a high school on a field trip. Most of the students died, but many survived because some teachers and crew members forfeited their place in the escape boats to stay behind and save others. Did they sacrifice their lives because they were trained to do so as teachers and ship crew?

I wonder, are these examples of just truly exceptional people placed in exceptional circumstances? In one way yes, but in another sense no. Although not all are covered as prominently in the news, events like these, in which people forfeit their self-interest or even their lives for complete strangers, are not hard to find at all. It turns out that helping others without a reward to oneself is not as rare as you might think.

Is helping others written in our DNA?

Among nonhuman animals, examples of self-sacrifice for the good of the group abound. Ants and bees are perhaps the best-known exhibitors of this behavior. Worker ants and worker bees toil all their lives and will fight to the death against invaders. Monkeys will also protect their companions when a menace to the group appears: they shout loudly so that the rest of the group can escape to safety. In doing so, they risk drawing the predator's attention to themselves. Are these ants, bees, and monkeys all just foolish? It would serve their individual interests best if they quietly escaped danger, so why don't they? What motivates organisms—all of which are fighting for evolutionary success—to forfeit their own self-interest to protect the group, if only briefly?

The well-known sociobiologist Edward O. Wilson, at Harvard University, looked for the answer among ants and bees. In ant and bee groups, the queen is solely responsible for reproduction. Ants and bees are all offspring from the same queen, and they may be genetically closer to each other than human siblings. In other words, there is small differentiation between individuals. The offspring are countless copies of one "self." Because an individual's genes live on even if it dies, individual sacrifice is not a bad deal from the gene's perspective. Therefore, all members of the group devote their lives to sustain the family, collectively taking care of the next generation.

When you expand your thinking beyond the individual level and consider the group as a whole, the self-sacrifice of eusocial animals such as ants or bees is fundamentally extremely selfish behavior. Even if I (as an ant or bee) die, the complete set of my genes will live on in the members of my group. Sociobiology, a new field based on this principle, became quite popular through books such as *Sociobiology* (1975) by Wilson, and *The Selfish Gene* (1976) by Richard Dawkins.

Monkeys also act for the benefit of the group, but in a different way. Monkeys do not share exact copies of their genes like ants and bees do. Instead, monkeys live with other members of their family, which share a significant amount of genetic material. If genes are shared, altruistic behavior that benefits the group can be beneficial to the self. This observation was first explained by William Hamilton and has come to be known as Hamilton's law. The law proposes that altruistic behavior is proportional to the amount of genetic material shared between relatives; altruism can therefore be calculated mathematically, like this:

$$rB > C$$

The equation states that altruism happens when the benefit for the recipient (B) multiplied by genetic relatedness (r) is larger than the cost for the altruist (C). For example, probabilistically speaking, I share ½ of my genes with my siblings, and ⅛ of my genes with my first cousins. If the benefit and the cost are the same, then 2 siblings and 8 cousins are equal in value. In other words, Hamilton's law says that, from a genetic point of view, it is a fair trade if 2 siblings or 8 cousins can be saved in exchange for my life. The same amount of genetic material will be saved in the next generation.

The specific numerical values in this calculation are less important than the concept that arises from it: genetic determinism, the view that the individual is only a receptacle for genes. This model explains why an adult man, after becoming a father, is ultimately serving his own self-interest by looking after the welfare of the offspring who share his genes. If this is true, though, how is altruistic behavior toward strangers explained? According to this model, sacrifice for the sake of strangers is just a residual habit from our long evolutionary history of living with relatives.

As we'll see shortly, however, cooperation within human societies cannot be explained in that way—partly because human families turn out not to be exclusively connected by genetic relatedness. The social relationships that we cherish tend to be with unrelated people who are "like family." In many cultures, "father" is a generalized term used by children for the man who is residing with the family but is not necessarily the biological father. Because there is no strict enforcement of monogamy in these cultures, this "father" and his "children" may not actually share any genes. Perhaps fatherhood is more clearly known in societies enforcing monogamy? Not necessarily. Many families are blended through divorces and remarriages, and many others have children through adoption, or through fertility clinics involving eggs or sperm of others. Modern society may be rewriting what it means to be a parent or a child, but this is not a completely new cultural change.

Human societies have always been much larger than the network defined by blood relationships. Think of the people you've communicated with so far today—through phone, email, social media, or in person. How many of them are your actual family or relatives? Most likely, much of your communication has been with people who share no genetic relatedness with you. Furthermore, you might not ever see many of them again. Even if you help these nonrelations, your genes reap no benefits at all.

Surprisingly, humans often risk their own lives for such "strangers." We give blood, donate money, and share food. We even donate our organs. Often, we do not ask for anything in return, sometimes even insisting on anonymity. This behavior is incredibly rare in nature.

The human family transcends blood relationships. We use kinship terms for our blood relations, but also for our friends and for relations through marriage. We may call an elderly woman "auntie" or an elderly man "uncle." We bring into our family network people who are not related by blood.

We do not do this out of nostalgia, nor is such behavior a remnant from a time past, when we used to live close to our relatives. Fictive kinship—treating and calling non-kin as if they were related—may not be a recent development; it may have always been with us. We take care of friends because friends are family too. This was true in the past, and it is true now. The concept of fictive kin—of connecting with unrelated "brothers" and "sisters," even to the point of giving up our self-interest for them—is indeed unique to humans.

1.8 million years of altruism

When did humans first start to exhibit such strange altruistic behavior? The first signs of altruism were found in Neanderthals, our extinct relatives. In the early twentieth century, a strange Neanderthal fossil specimen was found in La Chapelle-aux-Saints, France. This specimen had severely bent bones, leading people to think that Neanderthals might have had a stooped posture. They thought Neanderthals were not as smart as modern humans—a trait that they believed was implied by the sloppy stance of the Neanderthals. Similarly, the Neanderthal mandible (jawbone) jutted out, with a sunken mouth, which was interpreted as reflecting a stupefied appearance. The reconstructed appearance based on this La Chapelle specimen had a strong influence on the perceived image of Neanderthals for years to come.

Later research demonstrated that the bones of the La Chapelle Neanderthal were bent because of severe inflammation from arthritis due to old age. The sunken mouth was caused by missing teeth. The La Chapelle specimen had lost most of his teeth during life, perhaps from old age or for other reasons. When a tooth falls out after or near death, the space left remains unfilled. In contrast, if the individual lived beyond the tooth loss, the space in the gum is filled with new bone tissue and becomes smooth as a result of bone resorption. La Chapelle's mandible clearly showed that this individual had lost his teeth well before death. The space was filled, and the bone was smooth.

What can we conclude from this evidence? The La Chapelle individual appears to have suffered from arthritis due to old age, and lived well beyond losing most of his teeth. People started to call this specimen the "old man of La Chapelle," although "old man" from a Neanderthal's perspective probably meant between thirty and forty years of age, given the short and brutal lives of these hominins. The La Chapelle specimen raises an intriguing question: How could an "old man" Neanderthal with almost no teeth, who wasn't even fully mobile because of arthritis, survive through the Ice Age in deep, snow-covered mountains? It would have been extremely difficult to get food and to chew, even if he somehow managed to acquire a meal. There was only one way: someone must have been taking care of him. Paleoanthropologists concluded that La Chapelle could not have survived without the help of kin or neighbors.

Moreover, La Chapelle is not the first Neanderthal to show evidence of care. In the 1950s a Neanderthal fossil (Shanidar 1) discovered in Shanidar cave, Iraq, showed that a serious injury had been sustained during youth. Shanidar 1 was blind in his left eye, as clearly shown by his skull's morphology. The optic nerves from our eyes pass through a hole in the back of the bony plates that protect our eyes. In Shanidar 1's skull, that hole was filled in, indicating that his optic nerve was dead and did not pass through it. Furthermore, there was evidence of a big wound on the left side of his skull, which would have led to serious damage in the left hemisphere of his brain. As a result, he most likely was paralyzed on the right side of his body. His right arm was also shriveled and emaciated, and he likely walked with a limp. This fossil specimen also had traits of an older man, meaning that he probably sustained the serious injury when he was young, and someone (or some group) must have taken care of him into old age.

Recent discoveries show that this kind of altruistic care was seen in hominins much older than Neanderthals. Early hominins from Dmanisi, Georgia, are dated to be as old as 1.8 million years. Some of these early hominins show evidence of survival long after losing all their teeth. One of the Dmanisi fossils clearly lived to be an "old" person, its age estimated from the state of its cranial suture closures. These Dmanisi hominins also lived during the Ice Age, when it would have been difficult to acquire food. Unless someone brought food to them and processed it to the point that it could be consumed by an edentulous (toothless) person, they would not have lived long—except the fossils tell us that they did.

At the time of the Dmanisi hominins, the genus *Homo* had just appeared on the world stage. Hominins of this time were not too different from their earlier predecessors, *Australopithecus*. They had a similar small body size and modest brain capacity. The one difference was this evidence of altruism, the fossil record showing that members of the genus *Homo* took care of each other from the moment they appeared in the Paleolithic record.

Cooperation and altruism as powerful weapons

How did our ancestors come to help each other? Why would they have started showing altruism toward possible strangers? One reason might be the fact that our predecessors were small and weak; strength could be found only through social alliances. To adapt to environmental changes, humans became more versatile and cooperative, rather than physically robust.

The Ice Age was not consistently cold; the weather varied quite a bit. At times it was somewhat warm. Sometimes there were periods of drought; other times it rained for days. The landscape changed, as did sea levels. Islands became landmasses, and oceans gave way to mountains. When the climate changed, fauna and flora had to change with it or risk extinction.

To survive these dramatic environmental changes, ancestral humans had to be flexible. At some point they came to an important realization: changes in the environment did not immediately lead to a completely new set of environmental resources. In fact, the environment sometimes returned to circumstances that humans had experienced before. During those times, we could use archived knowledge from past experiences to our advantage. We essentially evolved to be more versatile by relying on our ability to store and transmit cultural information to the next generation.

And the best source for this stored information? The elderly. The elderly kept an archival memory of their life experiences and communicated those experiences to their kin. By inheriting and applying the information from previous generations, humans could now adapt to and live in environments where no ape had been able to live before.

At first, humans may have respected and helped older people as an invaluable source of information. But at some point in time, helping others took on a new form, as something unconditional and universal. Humans acquired a behavior that nonhuman animals do not have: universal cooperation and altruism. Forgoing self-interest for others, sharing resources with strangers, yielding the self for others, taking care of neighbors who cannot take care of themselves, and making a contribution to society are all part of human behavior. Without consciously trying, humans have been practicing "love your neighbor as yourself" for countless millennia.

This is a rather comforting thought, not simply for humanity as a whole, but also for me personally, because I am severely nearsighted. I know that even if I had lived in a society without glasses, whether among Neanderthals or earlier hominins, I would have stood a good chance of surviving. None of my fellow hominins (I hope!) would have left me to be eaten by a cave bear or starve.

EXTRA

NEANDERTHALS FROM THE SHANIDAR CAVE

The Shanidar cave mentioned in this chapter is in the mountainous region of Kurdistan, in northern Iraq. In the 1950s and 1960s, a team from Columbia University excavated several Neanderthal individuals with a wide range of ages. Among them, Shanidar 1 and Shanidar 4 are the best known. The individual mentioned in this chapter, Shanidar 1, is particularly well known because it shows signs of being cared for after a leg injury during youth until its death at approximately forty years of age.

The other individual, Shanidar 4, is well known because

of signs that it was intentionally buried. Pollen discovered in the soil surrounding the fossils appeared to be indicative of flowers that have historically been buried with the dead. This interpretation supported the idea that humans are inherently loving (the original flower people, so to speak). Recently, researchers have argued that the flower pollen was carried into the burial cave accidentally by natural elements, such as animals or wind. That conclusion, however, has no bearing on other Paleolithic evidence of purposeful burials and care, not just in Iraq, but from burials across different continents. The debate continues.

CHAPTER 13

King Kong

I think we can all agree that dragons are imaginary animals. Surprisingly, though, there was a point in time of human evolutionary history when dragons made an appearance (albeit briefly, and mainly because of confusion). What's more, they were thought to be relatives of humans, no less. This story is about *Gigantopithecus,* the largest primate ever, who was almost misclassified as a dragon.

In the early twentieth century, traditional apothecaries in China sold a nearly infinite number of medicinal products. Among them was something called "dragon bones," usually sold as a powder made from ground-up fossilized bone, but sometimes these dragon bones were sold intact. In Chinese medicine, dragon bones were extremely popular and in short supply.

Among the many Europeans who spent time in China in the early twentieth century was the German paleontologist Gustav Heinrich Ralph von Koenigswald. One day, while browsing in an apothecary in Hong Kong, von Koenigswald was shocked to see a dragon bone specimen on sale. As a paleontologist who had carefully studied animal bone morphology, von Koenigswald knew in an instant that the "dragon" fossil was a fossilized ape tooth. He also noted something very strange about this particular tooth: it was larger than any ape tooth he'd seen before. Von Koenigswald bought the "dragon bone," which turned out to be an ape's right third molar, and he published his findings in a 1952 paper, announcing a new fossil species, *Gigantopithecus blacki*. *Gigantopithecus* means "gigantic ape," and *blacki* honors the famous paleontologist Davidson Black.

It may be disappointing that the dragon bone did not belong to a dragon, but the real story is even more exciting: the tooth belonged to a "monster"-sized ape. *Gigantopithecus* would have been similar to a gorilla, but much, much bigger in body size. Unfortunately, even with the interest generated by von Koenigswald's publication, no other bones of the gigantic ape species have been found; they are known through teeth and mandibles (jawbones) only.

Was King Kong real?

The limestone caves in southern China have always been popular as agricultural land. While plowing this land, farmers unearthed hundreds of *Gigantopithecus* teeth throughout the 1950s and early 1960s. Paleontologists descended upon the site and published their findings. Despite focused efforts, however, the only specimens found have been three mandibles and more than a thousand teeth. No other body parts have been discovered so far. Even today, although papers continue to be published about new discoveries of *Gigantopithecus* fossils, the found specimens are still only teeth.

Wait, it is too early to be disappointed. As discussed earlier, paleoanthropologists and paleontologists can gather a lot of information about an individual from just mandibles and teeth. For example, body size can be estimated from the size of individual bones. At present, the primate with the biggest body is the gorilla, with males weighing as much as 400 pounds, and females as much as 200 pounds. In comparison, from the size of its teeth *Gigantopithecus* may have weighed two and a half times as much as a male gorilla, or 1,000 pounds, and may have been nearly 9 feet tall. Some scholars argue that *Gigantopithecus* weighed three times as much as a male gorilla! A more accurate estimation would require discovering skeletal parts that directly support body weight, such as arm or leg bones; but even with only its teeth, we know that the owner of these "dragon bones" had a gigantic body. Instead of a dragon, though, it was a King Kong.

Why do some animals grow to be so big? The easiest answer is found in male-male competition. Let's assume that only a few males have access to all fertile females. Males would have to engage in intense competition to be included in the chosen few; in such competition, a larger body would be an advantage. In the process of mate selection, females would prefer males with bigger bodies, and those males would get to transfer their genes to the next generation. The more intense the competition for access to fertile females, the bigger the body of the winning male has to be (or, as biological anthropologists would put it, the greater body size sexual dimorphism there must be).

Species with a high degree of body size sexual dimorphism tend to have a mating system with a single male and multiple females, with one male monopolizing the access to the females in the group. In contrast, species in which males and females do not have a big difference in body size tend to have single male–single female mating systems. In this case, because most males can mate with females, there is only weak competition among males, and both males and females tend to participate in taking care of offspring. Thus, from the difference in body size between males and females, we can make inferences about their mating system.

In the case of *Gigantopithecus*, however, male-male competition is not the reason for big body size. In 2009, I was at an international conference in China. I met Zhang Yinyun, a Chinese paleoanthropologist who had investigated *Gigantopithecus* for many years. Since he was retired, Zhang gave me the data he had collected on *Gigantopithecus* and asked me to continue his unfinished research. It turned out to be a big box of 3 × 5 index cards. Each card was filled with notes for every *Gigantopithecus* tooth that had been found. When I returned to the United States, I started to delve into the case.

The teeth were gigantic; that was to be expected. The difference in body size between males and females was also quite pronounced. This information we already knew. But one particular feature caught my eye: the canines. The canines were too small in comparison to the proposed body size and the rest of the teeth. The canine size difference between males and females was also negligible. This small difference was puzzling because canines are important in male-male competition. Even if overall body size does not differ greatly between males and females, a big difference in canine size still indicates intense male-male competition. Chimpanzees are a good example. Conversely, if male-male competition is weak, canine size is not different between males and females. In this case, humans are a good example: men and women do not show a significant difference in body size or canine size.

No matter how much males and females differ in body size, without a difference in canine size we can confidently assert that male-male competition among *Gigantopithecus* individuals was uncommon or nonexistent. If it wasn't male-male competition, though, what other factor could have led to the big body size of *Gigantopithecus*?

The reason may lie with predation. Having a big body obviously makes it easier to confront predators. Predators do not discriminate between males and females when it comes to prey, however, so this doesn't fully explain the body size *difference* between *Gigantopithecus* males and females. Instead, this difference likely has something to do with reproduction. In apes and other primates, body size increases as growth period lengthens. In other words, they grow bigger by growing for a longer period of time. A longer growth period, however, delays sexual maturation. Delayed maturation for females has particularly negative consequences for reproduction, since delaying pregnancy and childbirth is not necessarily advantageous, and quite often is disadvantageous. Therefore, females often have shorter growth periods, while males can afford to delay maturation and increase body size. Research on primates shows that for females, the length of the growth period tends to be stable for a species, with small individual variation. In contrast, males show variation in growth period and consequently in body size because of individual factors, as well as sensitivity to environmental changes.

This seemingly unassuming riddle of *Gigantopithecus*—the small canines, consistent between males and females, combined with astounding body size of the males—hints at the presence of a formidable predator in *Gigantopithecus*'s environment. But what kind of predator could possibly compel *Gigantopithecus* to have such a gigantic body? What sort of hunter created this "King Kong"? The answer, perhaps surprisingly, may be us. Or more accurately, our ancestors.

Man versus beast

Gigantopithecus lived in southern China from 1.2 million to 300,000 years ago. During that time, *Homo erectus* lived all over the Asian continent. *Homo erectus* hunted big game. In *Homo erectus* sites, such as Zhoukoudian, China, numerous horse and other animal bones were discovered. The horse bones appear to have been discarded after the animal was butchered and consumed. Some researchers even argue that horses became extinct in Asia because of overhunting by *Homo erectus*.

Is it possible that *Homo erectus* also hunted *Gigantopithecus* to extinction? Thus far, there are no data to support this idea. One would expect to find at least some sites with bones of both *Homo erectus* and *Gigantopithecus*, but such a discovery has never been made. Russ Ciochon, an anthropologist at the University of Iowa, once reported a discovery in Vietnam of *Homo erectus* teeth in proximity to *Gigantopithecus* teeth. Ciochon later rescinded that claim, when the hominin teeth turned out to belong not to a hominin, but to another ape.

This doesn't mean there was no relationship at all between *Homo erectus* and *Gigantopithecus*, however. Some paleoanthropologists posit that, even if the relationship wasn't necessarily that of the hunter and the hunted, there was still intense competition between these two species—competition that led to the extinction of *Gigantopithecus*. I agree. *Gigantopithecus* lived in a bamboo ecozone, in competition with the giant panda, whose main diet is bamboo. This competition intensified when *Homo erectus* entered the picture. On the surface, the uptick in competition is puzzling, since *Homo erectus* did not eat bamboo. Why did the competition intensify?

The answer lies in the possibility that *Homo erectus* used bamboo to make tools. The *Homo erectus* individuals of East Asia made stone tools that were crude in morphology and small in quantity, compared with what has been found in Europe or Africa. To explain this difference, scholars argue that the Asian *Homo erectus* must have used bamboo, a rich resource in Southeast Asia, instead of stones, to make sophisticated tools. Because bamboo tools are not preserved in the archaeological record like stone tools are, use of bamboo for tools would make it appear as if *Homo erectus* in Asia did not make many tools. Those who argue that the Asian *Homo erectus* did make tools out of bamboo, and perhaps even constructed shelters out of bamboo, suggest that this species caused deforestation by using too much bamboo. At the hands of *Homo erectus*, *Gigantopithecus* would have lost its habitat.

Furthermore, *Gigantopithecus* must have experienced food shortages. Although *Gigantopithecus* lived in bamboo forests, bamboo was not its main diet. Its teeth show that, like any other ape, it ingested a wide range of food. In particular, *Gigantopithecus* teeth had a high frequency of dental caries, implying a love for sweet and ripe fruits. They also showed a high frequency of enamel hypoplasia, a condition caused by malnutrition, particularly during growth periods. No matter how rich a tropical forest is with plant food, it clearly did not have enough food for the King Kong–sized *Gigantopithecus* to eat as much fruit as it must have wanted.

The Middle Pleistocene (the time period during which *Gigantopithecus* lived) increasingly fluctuated between warm-and-wet weather and cold-and-dry weather, gradually becoming colder and drier overall. The survival of *Gigantopithecus* was threatened by the cold and dry climate and the constant shrinking of its habitat. Add food shortages to the climate change and it is easy to see how *Gigantopithecus* had difficulty sustaining its big body, even without competition from others, hominin or not. Ultimately, the biggest primate ever found—*Gigantopithecus*—went extinct.

The story of *Gigantopithecus* is just one of many tragedies in biology. Ancestral humans were in competition with other animals over decreasing resources throughout the Pleistocene, and they came out as the most predominant species in the world, outcompeting every other organism in their way. *Gigantopithecus* could be just one of many large mammals that went extinct during the Pleistocene.

Every time I think of *Gigantopithecus,* I'm reminded of orangutans. Orangutans currently live in the tropical forest of Southeast Asia that *Gigantopithecus* once inhabited. Orangutans also have a large body and substantial body size differences between the two sexes. Orangutans, however, do not live in groups of a single male and multiple females, nor do they live in groups of a single male and a single female. Strangely enough, orangutans live solitary lives. This solitary behavior is unusual for primates, and some have suggested that the orangutan's solitary lifestyle is a desperate adaptation to survive in the face of encroachment by the most formidable predator on Earth: humans. Perhaps they learned this lesson from the extinction of *Gigantopithecus*.

In the near future, after deforestation and climate change potentially wipe all other apes off the map, humans might be the only primates left. At that point, we may not be so proud of ourselves.

EXTRA
ROMANCE OF THE GIANTS

Throughout history, many human cultures have featured mythologies of humanlike giants. Goliath of the Judeo-Christian Bible is only one of the numerous giants that appear in the legends of Scandinavia, Rome, Greece, and elsewhere. Even now, we hear of people who claim they saw Yeti, the Himalayan giant, or Sasquatch, the North American giant. Some people even devote their lives to the search for such giants.

Intriguingly, in the history of paleoanthropology, a giant hominin also once made an appearance. Among the hominin fossils discovered on the Indonesian island of Java were fragments of skulls, mandibles, and teeth of a startlingly large size. They were given the name *Meganthropus*, meaning "big humans." Franz Weidenreich, famous for his associated research on Peking Man, wrote a book titled *Apes, Giants, and Man* in 1946 about the unusual Java find. *Meganthropus* is now considered to be a part of the *Homo erectus* species, and the existence of a giant human species is not even considered a remote possibility.

Gigantopithecus, however, was truly an ape species of remarkable body size that lived at the same time, in the same region, as *Homo erectus*. I wonder whether the existence of *Gigantopithecus* somehow remains so firmly embedded in the collective memory of hominins that it is the foundation for our universal fascination with, and mythology about, giants. This is all just speculation—yet another intriguing path of human evolution that inspires our imagination.

CHAPTER 14

Breaking Back

We think of ourselves as the smart species. Our traditions of learning and transmitting knowledge to the next generation rely on our smartness. We tend to view our "excellent brains" as the principal trait that sets humans apart from other animals.

When we talk about excellent brains in the fossil record, we really mean big brains. Humans belong to a group of animals with dramatically increased brain size. If you look at early-hominin fossils, the brain size is barely 450 cubic centimeters (cc), similar to that of a chimpanzee adult. This is about one-third the size of a modern human adult brain. The hominin brain had almost doubled in size, to 900 cc, by 2 million years ago, and by 10,000 years ago it approached the modern average of 1,400 cc.

What led to this change? Stone tools might have; they start to appear in the archaeological record about 3–2.5 million years ago. Although language does not become fossilized, we assume that language must have appeared after the big brain appeared. The increase in brain size was thus possibly the first uniquely human trait to arise. And that trait is reflected in the name of our species: *Homo sapiens*, meaning "knowledgeable human." It seems we have a natural propensity to seek knowledge, to learn about and manipulate our environment, and to develop tools and technology to enhance our abilities.

Yet cerebral capacity may not be the most important mover in human evolution. It's clearly *not* the first uniquely human trait that appeared in our evolution. The earliest trait that gave us our "human-like" appearance is, in fact, found at the opposite end of the body: our feet (see Chapter 3). The use of our feet for upright walking, however, came at a cost to our backs, which can quite literally break from the stress of upright locomotion.

Humanity: all legs and no brains

In 1974, Donald Johanson, then at Case Western Reserve University (now at the Institute of Human Origins), found a hominin fossil at an excavation in Ethiopia, East Africa. Back at the camp on the night of the discovery, the radio was playing the Beatles' "Lucy in the Sky with Diamonds." Inspired by the song, the excavation team named the new fossil "Lucy." It was the moment of birth for the most famous fossil in the history of anthropology.

Lucy belonged to a species that lived about 3.3 million years ago: *Australopithecus afarensis*. While many australopithecine fossils were discovered in the 1970s, Lucy was the face of the species, so to speak, even though she lacked an actual cranium. At the time of her discovery, Lucy was the oldest known hominin fossil. And since this specimen was missing a head, researchers were unable to answer the public's most pressing question: How big was the brain? This question was answered by other fossil specimens of *Australopithecus afarensis*, discovered later: Lucy's kind did not have a big brain. Paleoanthropologists were focused on another trait entirely, the legs.

Skeletal elements that are preserved as fossils can tell us whether an animal from the past walked on two feet or four. Animals that walk on four legs have their body weight distributed evenly among all four limbs. But animals that walk on two legs have their body weight resting on only the two legs, while the arms do not show signs of bearing body weight. The joints that bear body weight tend to be bigger, so we can see from the size and shape of a joint whether it was responsible for weight-bearing locomotion. By comparing the size of the hip joints, where the legs are connected to the body, to the size of the shoulder joints, where the arms are connected to the body, we can determine how many limbs a species used to move about.

On examining the shoulder joint of *Australopithecus afarensis*, paleoanthropologists found it to be small. This meant that the shoulder did not support the body weight. In contrast, the hip and knee joints were larger, indicating that they were load-bearing. The joint shapes also appeared different from those of other primates. The knee joint was flat and sturdy for stable support without much range of motion; other primates have a round knee joint to maximize range of motion. And the hip joint was fitted deeply to maximize stability, so that it couldn't easily be dislocated. The shoulder joint did not have these morphological adaptations. All of this evidence supports the hypothesis that the body weight of our ancestors was supported by two legs.

Walking on two legs is quite different from standing on two legs. Stand up and try walking right now. With every step you take, notice that only one foot is in contact with the ground as you switch between feet. That one foot—and finally only one point on the foot, the big toe—takes the whole body weight. What we describe as walking on two feet is, in fact, walking on one foot at a time. The biggest problem with walking on one foot is the risk of falling from the loss of balance as each foot takes turns supporting the body weight. To solve this problem, humans underwent morphological changes in the toes, ankles, knees, legs, and pelvis. The muscle connectivity of the pelvis and femur were also repurposed for functional loading on two legs. The hip and thigh muscles that had once been used to propel the leg forward now took on a new function of stabilizing the side-to-side swaying of the body.

The body weight supported by one leg is transmitted to the front of the foot (big toe) just before shifting to the other leg and foot. Because the foot has to briefly bear the whole weight of the body, the human big toe is the biggest and sturdiest among all the toes, and it aligns with the other toes, facing forward. This arrangement is quite different from that in other apes, whose big toes face sideways like our thumbs.

Owen Lovejoy* of Kent State University and Tim White of UC Berkeley argued that Lucy walked on two feet. Moreover, previously, in 1979, paleoanthropologist Mary Leakey had discovered a volcanic ash site in Laetoli, Tanzania, where a group of footprints were clearly marked on top of the ash layer. Dated to be 3.6 million years old, this find was indisputable evidence of bipedalism, but the field of paleoanthropology started on a long debate about what it meant for our ancestors to be bipedal. Discovery of Lucy only added fuel to the fire.

Paleoanthropologists continued to debate for the next twenty years whether hominins had big brains or bipedalism first. It was quite difficult to entertain the idea that "humanness" could have started with our feet and not our brain. The debate is now settled, and it is accepted that bipedalism was the first of the two traits to appear in human evolution.

* This is the very same Lovejoy who made an appearance in Chapter 2, where I discussed the Lovejoy model of human origins that was published in the journal *Science* in 1981.

Bipedalism, the source of back pain

Even after the shift to bipedalism, the road to becoming human was strewn with obstacles. There was a price to pay for walking on two legs. Walking upright means that the body needs to stay upright. As a result, a significant proportion of the body's weight is concentrated in the lower back (consisting of the lumbar vertebrae and pelvis). Body weight is then distributed to two legs; during walking, it is carried by one leg at a time. Consequently, the lower back, hip joints, and knees all have to transfer and carry the body weight constantly, making humans particularly prone to pain in the lower back and knee joints. Animals that can evenly spread their body weight between front and hind legs do not have this problem.

Moreover, females are faced with the even greater burden of carrying the additional weight of a child. Until recently in human evolution, women often spent most of their adult life either pregnant or nursing a child. They started the cycle of pregnancy and childcare as soon as they reached maturation, giving birth to five or six—sometimes as many as twelve or more—children throughout their lifetime. After menopause, these women became grandmothers and began to assist in the care of their grandchildren. For human females, this lifetime burden on the lower back and legs was quite painful and damaging over time.

All of this carrying and lifting extra weight also burdened the heart. In animals that walk on four legs, the heart is relatively high in the body, allowing it to use gravity when sending blood to the whole system. (With a neck 2 meters long, giraffes are a notable exception; they adapted by having a very small head and a very large heart in proportion to their body size.) The human heart is located relatively low in relation to the layout of the whole body; it's closer to the vertical midpoint of our body. As a result, the chest, shoulders, arms, and cranium (with its big brain) are all positioned higher than the heart.

This schema means that the heart must pump a substantial amount of blood against gravity, something that the mammalian heart of the past was not exactly designed to do. Moreover, in humans the heart not only is located lower in the body, but also has to supply blood to a much enlarged organ, the brain. The human brain is quite literally a monstrosity, a gigantic organ that consumes a lot of energy and needs a lot of blood. The adult human brain at rest consumes 20–30 percent of total daily energy intake, and for children the consumption spikes to as much as 50–60 percent. The magnitude of this problem dwarfs the case of the giraffe.

The body part to which the heart must pump the greatest amount of blood is also the highest point in the body, making the heart's job relentlessly difficult. Thus the heart is like Sisyphus of the Greek mythology, burdened with a never-ending task. It is no wonder that the heart seems to want to call it quits at any moment. In this light, it is perhaps inevitable that humans have higher heart-related mortality compared to other animals.

Human civilization in exchange for back pain

Walking on two feet wasn't all pain and no gain. Thanks to bipedalism, humanity gained another uniquely human trait: toolmaking. Bipedalism freed our hands and arms from locomotion. The free hands and arms could now make and use tools.

Furthermore, once the upper body was freed from locomotion, the diaphragm was freed as well. Breathing was no longer constricted by movement, so vocalization became possible. And vocalization enabled language. In this way, the foundation of human culture and civilization—tools and language—was complete.

One could also say that bigger brains are a result of bipedalism—perhaps the greatest achievement of bipedalism. In order to make and use tools, advanced cognition is also necessary, which is associated with bigger brains. The brain, however, cannot simply become big on its own. The brain is an organ made of fat, which requires a diet high in calories, in the form of fat and protein. This kind of diet is possible only with a strategy that allows for the regular procurement and consumption of high-quality food (meat) using the technological innovation of tools. All these interrelated components could be possible only after bipedalism was established.

While walking on two feet, humanity could dream of culture and civilization, but at the cost of back pain, heart disease, and dangerous childbirth. Take a moment to stand up from your seat and stretch for your heart and lower back, our sacrifices for humanity. And while you're at it, how about a thank-you message to your mother? Use one of those tools of civilization—whether a phone call, a text message, or a post on the web—that we acquired in exchange for the mother's pain of childbirth.

EXTRA

CHIMPANZEES CAN WALK UPRIGHT TOO!

Obviously, humans are not the only animals that can walk on two feet. Gorillas, chimpanzees, some lizards, and birds walk on two feet too, although in a different way. But the thing is that these animals have other ways to move about. For example, many birds can fly. Flightless birds can do other things: penguins swim underwater for a very long time; ostriches use their two feet to move, but they can run as fast as 35–45 miles per hour. Chimpanzees and gorillas walk equally well on four feet, and they also climb trees and brachiate through the trees at high speed.

Compare these examples to humans, who can only walk on two feet and have no other way to move about. Unlike ostriches, we cannot run for a long time, or particularly fast, without intensive training and talent. We might attempt to move on all fours for a short distance like a gorilla does, but this method of travel is awkward, inefficient, and cumbersome for an adult human. It doesn't take long for most people to give up moving on all fours and go back to standing up and walking on two feet. Any other way of moving is just not a reliable mode of locomotion for us. It's odd to realize that, in a way, this dependence on one limited mode of locomotion has led to some amazing evolutionary breakthroughs.

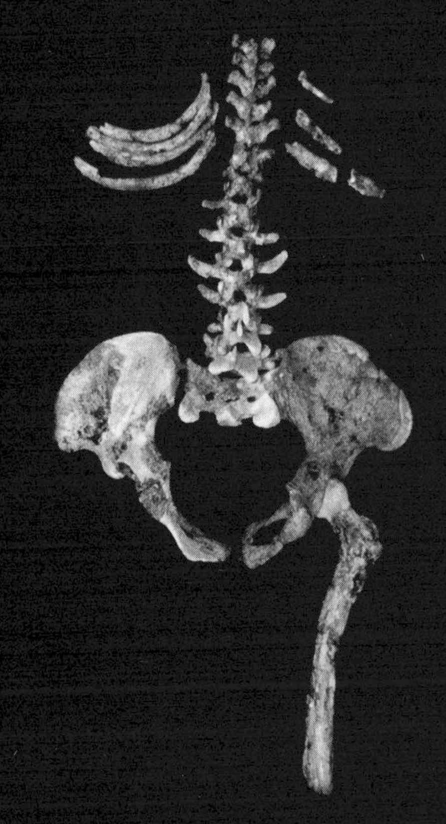

Fossil vertebrae (spine), ribs, hip bones, and femur (thigh bone) of *Australopithecus africanus*, discovered in Sterkfontein, South Africa. (© Milford Wolpoff)

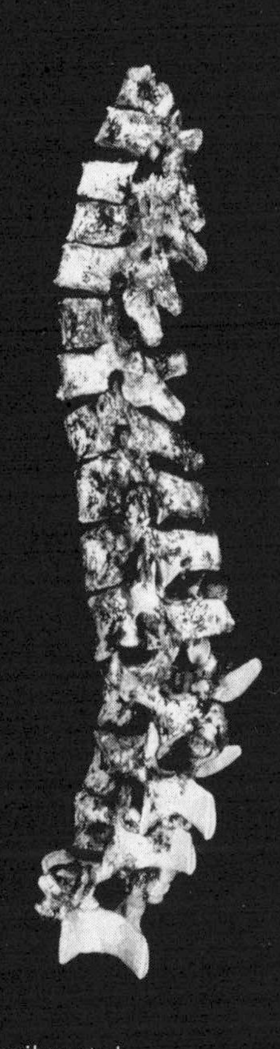

Fossil vertebrae (spine) bones of *Australopithecus africanus*, discovered in Sterkfontein, South Africa. (© Milford Wolpoff)

CHAPTER 15

In Search of the Most Humanlike Face

In early August of 2012, a side profile of the face of an ancestral hominin fossil was featured on the cover of the British scientific journal *Nature*; it was the face of *Homo rudolfensis*. It was rather long but clearly looked familiar and almost humanlike. Naturally, its appearance prompts the question, What makes a face "humanlike" anyway?

The hominin fossil that graced the cover of *Nature* has a unique ID: KNM-EM 62000. It was discovered in Koobi Fora, a renowned paleoanthropological site in northern Kenya, and was featured in several major news outlets, including the *New York Times*. What was such a big deal about this fossil that it was rewriting our evolutionary history? What made this specimen belong to the human family?

The story of "humanlike" faces cannot be told without also telling the story of the most famous family in the field of anthropology: the Leakeys. The Leakey family has devoted three generations to uncovering hominin fossils, from father and mother, to son and daughter-in-law, and finally to granddaughter. In a way, they are almost like a royal family of paleoanthropology. Their story is more than a simple family history; it is a record of a difficult fifty-year pursuit of a humanlike face and its origin.

Tracking the first hominin fossil

It all began in the 1950s. A young couple, Louis and Mary Leakey, were gaining attention through their excavation efforts in Kenya. The Leakeys, both originally from Great Britain, were seeking the fossil that would illuminate the origin of the genus *Homo*, a group of species that includes humans. In other words, they were looking for the earliest fossil evidence of our specific lineage.

At the time, there were two prominent fossil hominin species that could be relevant to the origins of the genus *Homo*. The first was *Australopithecus africanus* from South Africa. This species was thought to have appeared 2 million years ago. The second species, *Homo erectus* from East and Southeast Asia, was thought to have appeared 700,000 years ago. Given these dates, Louis and Mary Leakey surmised that the first *Homo* specimen appeared later than *Australopithecus africanus* but earlier than *Homo erectus*.

The Leakeys' brilliant achievements and success in paleoanthropology throughout the 1960s can be attributed to their persistent excavations. They demonstrated that early hominin ancestors thrived not only in southern Africa, but in eastern Africa as well. Mary Leakey's discovery of the Laetoli footprints in Tanzania provided valuable supporting evidence that bipedalism occurred in the human lineage 3.3 million years ago, long before the appearance of *Homo erectus*. The Leakeys also discovered the skull of the "Nutcracker Man," *Zinjanthropus boisei*, notable for its huge teeth and the pronounced sagittal crest at the top of its skull. Each successive discovery by the Leakeys was considered groundbreaking and a defining moment in the history of paleoanthropology.

The Leakeys found the quest closest to their hearts—to discover the origin of the genus *Homo*—not so easy. They were quite disappointed for many years, until they uncovered, in an excavation lasting from 1960 to 1963, a set of hand bones in the Olduvai Gorge, Tanzania.

The finger bones were delicate, tantalizingly suggestive of a toolmaking hand. The Leakeys wondered whether they had finally found the first ancestor of the genus *Homo*; perhaps their dream had finally come true. They named the specimen *Homo habilis*, meaning "handy human."

A father's dream, a son's triumph

The Leakeys' dream did not come true all at once. By themselves, the hand bones discovered in Olduvai Gorge did not constitute complete and perfect evidence of tool use, or of the genus *Homo*. New hominin fossil species are most often classified by skull morphology. If meticulous examination reveals a set of novel characteristics in the face and cranium, the skull is defined as a new species. To reflect a brain big enough to make useful tools, a large skull with a vertical forehead was sought. Fortunately, more *Homo habilis* fossils started to be discovered in subsequent excavations; yet, paradoxically, our understanding of *Homo habilis* became only murkier.

After *Homo habilis*'s discovery, hominin fossils from East Africa began to be classified into two species according to skull size. Small skulls were classified as *Zinjanthropus boisei* (the genus *Zinjanthropus* was later subsumed under *Australopithecus* or *Paranthropus*), large ones as *Homo habilis*. But there was a problem with this sorting. Skulls were discovered as fragments; even the most experienced paleoanthropologists often had difficulty accurately projecting the size and morphology of the whole skull. So, what did they do? They simply made an informed guess that if it "looked" big enough, it was *Homo habilis*. This imprecise approach caused doubts within the paleoanthropological community, and many debated the existence of *Homo habilis* as a single species.

In the midst of classification in the mid-1970s, an almost complete fossil skull was finally discovered at Koobi Fora. This new discovery was different from any of the fragmentary discoveries that preceded it. Assigned the catalogue number KNM-ER 1470, this new find showed characteristically *Homo habilis* traits. The young paleoanthropologist who discovered it was Richard Leakey. It was no coincidence that his name was Leakey, as he was the son of Louis and Mary. Richard grew up in excavation field sites with his parents, and by the 1970s he had followed in his parents' footsteps by becoming a respected paleoanthropologist himself.

The hominin fossil discovered by Richard Leakey had a big brain as people had predicted, with a straight and vertical forehead. The cranium was highly qualified to be a candidate for the ancestor of the genus *Homo*. Finally, it seemed as if *Homo habilis*, having started out as just a set of hand bones, was finally getting a humanlike face and becoming acknowledged as the earliest member of the genus *Homo*. But would this face be accepted by the scientific community without question?

Further findings began to complicate things. After Richard Leakey's discovery, hominin fossils continued to be discovered in East Africa. And they continued to be classified as either *Zinjanthropus boisei* or *Homo habilis*, depending on an increasingly vague and confusing standard of estimated size or shape. The standard itself was questioned from time to time. Some argued that for a species to make tools, the brain not only has to be big, but also must have traits associated with intelligence. Specifically, they argued that a high forehead in *Homo habilis* was necessary to confirm tool use, claiming that forehead angles were the best diagnostic for *Homo habilis*.

You might be able to imagine the confusion these differing methods of classifying *Homo habilis* caused in the field. Finally, paleoanthropologists realized that when all the samples of fossils classified as *Homo habilis* were compared, they showed a range of variation that was too great for them to be considered a single species. *Homo habilis* was a "hominin of a thousand faces," an arbitrary group without a cohesive definition.

At its core, the problem of *Homo habilis* has direct relevance to one of the most fundamental problems in paleoanthropology: variation. Let's imagine we're comparing two people. No two people look exactly the same; even twins look slightly different from one another. A random pair of people can vary in body size, sex, age, and countless other characteristics, but even with all this variation, we know that all people belong to a single human species called *Homo sapiens*. The variation that we see between members of the same species is called "intraspecific variation." In contrast to intraspecific variation, there is in*ter*specific variation, meaning differences between different species. Interspecific differences (such as between a person and a chimpanzee) are bigger than intraspecific differences (such as those between individual humans).

Now let's turn our example around. Starting with the pattern of variation, we can determine whether two individuals belong to the same or different species. For example, if two individuals look pretty much similar, differing only in size, they are likely of the same species. Likewise, if they show differences only between different sexes, they likely belong to the same species.

As paleoanthropologists were confronted with the growing number of *Homo habilis* specimens, they were surprised at the range of variation observed within the group. Should these excavated fossils of *Homo habilis* be considered different variations of the same species (intraspecific), or should they be classified as different species altogether (showing interspecific variation)? Some researchers argued that all of them did, indeed, belong to a single species. Others argued that at least two different species were represented in the group and the fossils should be reclassified—the sooner the better. The researchers who argued for a reclassification put the fossils with the biggest skulls into a separate species and called it *Homo rudolfensis*.

Through this process, the first complete *Homo habilis* skull, KNM-ER 1470, lay at the nexus of this dilemma. If based simply on skull size, the classification should have been *Homo rudolfensis,* not *Homo habilis.* Yet classifying KNM-ER 1470 as *Homo rudolfensis* presented a new problem: the fossil has a large skull, but the face looks exactly like those of the other *Homo habilis* specimens. This combination of skull and facial characteristics raised yet another possible explanation: that there must be a third fossil species, different from both *Homo habilis* and *Homo rudolfensis.* This solution presented the biggest problem yet, since no other fossil specimen found thus far looked like the one discovered by Richard Leakey. Scholars scratched their heads as they pondered the meaning of this fossil, an unusual and exceptional specimen.

Meave and Louise Leakey: discoverers of another early hominin

After his discovery of *Homo habilis,* Richard Leakey moved away from the excavation field and became more involved with the wildlife protection movement and political activism in the 1990s. He became especially involved in protection of the rhinoceros and retired from excavation altogether. But this was not the end of excavations by the Leakey family.

In 2008–9, another fossil skull—a face and mandible—was discovered in Koobi Fora, Kenya. This fossil, KNM-ER 62000, was the cover story of the August 9, 2012, issue of *Nature.* The point of the story was rather simple: KNM-ER 62000 looked quite similar to KNM-ER 1470, the fossil reported back in the 1970s by Richard Leakey. Richard's fossil was no longer "unusual and exceptional." KNM-ER 62000 was categorized as representative of the fossil species *Homo rudolfensis.* It showed that the early *Homo* genus did not have one species only. Instead, the possibility that at least two *Homo* species coexisted in Africa 2 million years ago became more plausible.

We've entered a new era in the forty-year-old debate about the origin and evolution of the genus *Homo*. But one thing has remained constant throughout: the name of the scientists who have led excavation and research efforts. KNM-ER 62000 was discovered by none other than Meave and Louise Leakey, the wife and daughter of Richard Leakey. These two women are continuing the second and third generation of the Leakey legacy by rewriting the story of the fossil species deeply associated with their husband and father.

EXTRA

YOU SEE WHAT YOU WANT TO SEE

KNM-ER 1470, the fossil find reported by Richard Leakey, has a history of ups and downs. It was first classified as *Homo habilis*; then, forty years later, it was reclassified as *Homo rudolfensis*. Name change is not a big deal in paleoanthropology, but this fossil got a face-lift as well. Scholars were intrigued by how closely—with its straight, vertical forehead and straight, vertical face—it resembled "modern humans." In contrast, movies and cartoons show "primitive men" with jutting mouths and brows, and flat, receding foreheads.

Yet Richard Leakey's fossil had a critical weakness that not many realized: its facial reconstruction was hypothetical. When the fossil was first excavated, it was broken between the nose and the forehead, and no connecting piece was found among the remains. Therefore, the angle at which the forehead meets the nose (and the rest of the face) had to be imagined by scholars. What if the fossil had been reconstructed differently, with the forehead receding more and the face jutting more forward? That reconstruction yields an equally likely hominin face! In other words, there is no compelling reason to reconstruct the forehead in a vertical, straight line, understood to be a telling characteristic of *Homo habilis*. Perhaps, the "humanlike" face of this fossil was a product of our own desire to have *Homo habilis* look more like an ancestor of humans.

CHAPTER 16

Our Changing Brains

My fifth-grade teacher once told our class, "Humans barely use 10 percent of their brains' processing capabilities. The rest is never used until the moment you die." This statement made me very sad as a child. Many years later the trailer of the movie *Lucy,* from 2014, began with the same declaration of brain underuse. That this huge brain of ours is left mostly unused seems tragic. Fortunately, this claim of brain underuse is false, lacking any supportive evidence.

Another common but false statement about the brain is that it hardens as we grow old. People often say the brain is "plastic" and flexible while we're growing up, capable of learning many things, but as we become adults, the brain settles into a fixed structure and becomes seemingly incapable of learning anything new. Research has shown that this statement also has no merit—another comforting thought for those of us hoping to continue to learn new things well into adulthood.

Adult brains and child brains are different

Even though the idea that older people cannot learn new things is false, clearly tasks that come easily to an elderly person are quite different from those that a child finds easy. For example, mechanical memorization is much easier during childhood. But pulling different pieces of information together, connecting them, and synthesizing that information is easier for adults than for children. Some of these differences come from the way the brain cells are organized. Children start making brand-new brain cells when they are fetuses, and they continue to make new brain cells through their growth periods. Having more brain cells means a higher storage capacity. By the age of six or seven, a child's brain is 80–90 percent of its adult size. After that point, fewer new brain cells are made, and the ability to retain new information becomes more difficult.

You might wonder, then, is the brain left on cruise control for the rest of its life? Not really. After early childhood, the brain starts the new task of making *connections* between brain cells, instead of simply making new brain cells. Connecting brain cells is not an easy or a simple task. Consider this: When there are two brain cells, there is one possible connection to be made between them. And when there are three brain cells, there are three possible connections. But when there are four cells, there are six possible connections. With six cells, the number of possible connections jumps to fifteen possibilities. In this manner, the number of possible connections increases exponentially as the number of brain cells increases. And given that the human brain may have as many as 100 billion brain cells, the number of possible connections between all possible pairs of brain cells is absolutely mind-blowing.

To be fair, no brain cell makes connections with all the other brain cells. A single brain cell may connect to only the several brain cells in its neighborhood or region of the brain. Even then, the number of connections is astronomically huge. Generally, a single cubic centimeter of brain volume is said to contain 600 million brain cell connections (called "synapses"), which means that a 1,400-cc brain has 840 billion synapses!

Synapses are important for collecting information and connecting the dots to draw a big picture. Even after new brain cells are no longer made, the brain is still quite active through synapses. This also means that the brain doesn't "harden," nor do we ever use only 10 percent of its capacity.

At any given moment the active part of a living human brain might be infinitesimally small, but we need the big brain to store important information. According to the "social brain hypothesis" proposed by Robin Dunbar, a psychologist at Oxford University, as our social group size increases, the information about the group members and about the relationships between them also increases astronomically. The brain stores and uses all this social information in various ways, similar to the way computers store and process large amounts of information, even if perhaps we're only playing solitaire at a given moment.

While human brain size is almost complete at the age of six to seven years, attaining adult brain size does not mean that the brain is functioning like an adult brain. Only after the brain reaches adult size does the real growth start. Knowledge and wisdom grow with the astronomically large number of synapses that we continue to make.

Humanity's big, social brains

When did the human brain first become big, equipped with the ability to store and process enormous amounts of information? For paleoanthropologists, this is an interesting question because if we know when, then we might know *why* the brain became big. In research about human brain capacity, 450 cc (about the size of a softball) is an important starting point. The earliest human ancestors, the genus *Australopithecus*, had a brain barely larger than 450 cc. A human newborn baby also has a brain barely bigger than 450 cc. Chimpanzees, the ape species closest in relation to humans, also have an adult brain size of about 450 cc.

Several million years ago, early hominins also had a small brain like the chimpanzee's, but by 2 million years ago the brain had doubled in size, to 900 cc. It is not a coincidence that 2 million years ago is roughly when our direct lineage, the genus *Homo*, first appeared. Then later, half a million years ago, the brain tripled its original size, to 1,350 cc. This trend of increasing brain size continued until 50,000 years ago, when Neanderthals achieved cranial capacities—1,600 cc—larger than the average brain size of modern humans today, 1,400 cc.

Why did the brain become bigger? The mainstream argument for many years has been that the drive for tool use is what caused the increase in brain size. As stone tools appeared along with the genus *Homo* 2 million years ago, a regular diet of meat and fat became possible through the use of artificially made tools to scavenge and hunt animals. The newly possible diet meant that brains could become even larger. For a long time, scholars argued that making and using tools, hunting, and advances in cranial capacity were the drivers of evolution in humans, but there's a problem with this theory: toolmaking doesn't take up much brain real estate. If our large human intellect was solely for making and using tools, then the human brain wouldn't need to be so big. This theory would mean that the human brain is unnecessarily, ridiculously big, not only in terms of absolute size, but also in terms of the particularly large cerebral cortex, known to be responsible for making high-level executive decisions.

Others, however, argue that such a highly advanced brain came about through tasks other than making and using tools. This is Robin Dunbar's social brain hypothesis, mentioned earlier. The cerebral cortex seems to be bigger in animals living in groups, and even bigger among large groups. Dunbar conducted research by eavesdropping on others' conversations for several years. He discovered that people talked mostly about other people rather than about religion, philosophy, or politics. This behavior was true regardless of the gender or social position of the conversationalists. We think that gossip is exclusive to women and that men do not gab, or at least they shouldn't. But if we listen closely, we find that men also love to gossip, and they discuss everyday mundane social matters. It seems that all people—men and women—simply love to talk. Dunbar argued that our "social brain" was used mainly for this kind of chatter—about the self or about things that happened to or between other people.

Indeed, it's true that social animals have bigger brains, such as dolphins or elephants. As group size increases, information about each member increases, and the number of relationships that each individual can form (and must maintain) also increases astronomically. Our brain pulls together all this information, stores it, and accesses it again when the need arises.

Early hominins certainly needed to form groups to survive. Hominins walking upright in the middle of the African savanna were rather pathetic in their ability to protect themselves, even when armed with hand axes. Compared to other predators, humans are relatively weak and cannot hunt very successfully by themselves. Instead, early hominins had to rely on hunting in groups, which in turn required a robust social structure. Furthermore, to survive in the ever-changing, cyclical, glacial environment, it was absolutely necessary to make and keep group connections to share information, and absolutely necessary to have a big brain to store that vital information. Social life for humans was a desperate means to survive, more than a hobby to fill leisure time. To collect and exchange information, and to understand it, advanced language developed as a means of communication. And chatting became the primary function of language.

The information exchanged through chatting is not necessarily directly helpful to everyday life. It certainly can be, though; in Africa during the dry season, observations such as "a temporary stream appeared in such and such place" or "I saw a lion catching a gazelle a little while ago, so let's go get the meat before the hyenas come" are critical to survival. But other forms of communication are less obviously connected to everyday survival, such as chatting about little things of daily life: "someone is having a baby" or "so-and-so is hanging out with you-know-who these days" or even something about one's health, like "my eyes hurt."

Some cognitive researchers have gone so far as to say that chatting is like grooming, but using the mouth. Most primates groom each other, and while picking out dirt and other stuff, they build friendships and bond. Grooming is an important aspect of their social lives. If I (as a monkey) came across another monkey that held a rank higher than mine, I would groom that monkey first, to show that I understood the social differences in our ranks. As group size increases (as it did in human evolution) and social relationships become more complex, though, it becomes impractical to directly groom every individual. Instead, we use words. In general, grooming is a singular, one-to-one task, at a given time. Using words, however, we can reach several individuals at once. Witness the birth of lip service.

Big brains need lean faces

The enlarged human brain did not come free of charge. Humans had to secure a great amount of energy for it, supplied in the form of animal fat and protein. For this, human ancestors had to scavenge carcasses left behind by other predators during the daytime, when the predators were asleep (see Chapter 5).

Except, securing meat does not directly lead to big brains. First some problems need to be solved. For one thing, other organs compete with the brain for limited energy sources. The digestive system is a good example. A finite energy supply dictates that the digestive system and the brain cannot both be big at the same time; for the brain to secure more energy to become bigger, the digestive system must get less energy. This is the basic idea behind the "expensive tissue hypothesis" proposed by anthropologists Leslie Aiello and Peter Wheeler at the University of London. Their studies of a variety of animals showed that brain size and digestive system size are inversely related.

The second problem is that the skull has to get bigger as the brain gets bigger. But for the skull to get bigger, the muscles holding down the bones that make up the skull have to become smaller first, to allow the skull to grow unencumbered. The biggest muscle connected to the skull is the masticatory (chewing) muscle. In other words, for the brain to get bigger, the masticatory muscles have to get smaller. Interestingly, a paper published in 2004 featured an experiment showing that a mutation in a gene (named MYH16) to make the chewing muscle small led to an enormously big skull in mice, making this hypothesis more plausible.

Let's look at how human evolution solved these two problems. Two million years ago there were three hominin species, with three completely different adaptations for the changing African environment: the vegetarian *Australopithecus/Paranthropus boisei*, the carcass scavenger *Homo habilis*, and the hunter *Homo erectus*. Of these, the vegetarian *Australopithecus/Paranthropus boisei* had a small brain (500 cc) but very big teeth. In contrast, the meat-eating *Homo erectus* had a relatively big brain (1,000 cc) but small teeth with small chewing muscles. And the brain size of the carcass-scavenging *Homo habilis* was at a midpoint between the other two, about 650 cc. There does, indeed, seem to be a surprising relationship between diet and brain size.

With big brains, as I've mentioned, hunting and gathering became a necessity. So, humans had to remember and synthesize information about patterns of animal migration, and about the continuously changing environment. In the process, the most important "weapon" in the human lineage appears to have been social cooperation. As the group size grew, the information about group members and the complex relationships between and among them increased immensely.

The big brains of humans were used for storing this enormous amount of social information and designed to access it at the right moment. This is the real reason for humans to have big brains. Even if not all of our brain cells are used at the same time, it is advantageous to have many brain cells and synaptic connections in our big brains. Having a big brain means being able to adapt to a rapidly changing environment, by using the massive storage of information that we can access to respond to a huge variety of environmental changes.

I will close this chapter by sharing a question with you. According to a publication that I coauthored with Milford Wolpoff at the University of Michigan, hominin brains consistently and gradually increased from 2 million years ago until 50,000 years ago. In this study, any data more recent than 50,000 years ago were not included. In essence, our research for this publication did not include recent changes in brain capacities.

So, what happened to brain size in the last 50,000 years? Have human brains continued to grow bigger and bigger? In actuality, the opposite might be true. Human brains might be getting smaller. This hypothesis needs to be studied in more detail with more data, but if supported, it is indeed an interesting proposition. If the human brain is getting smaller, the question is why. Is it that, after the invention of writing and advances in computing, machines have now taken over many of the tasks previously performed by our brains?

Perhaps modern humans are now experiencing a dramatic reversal of the direction of human evolutionary history over the past few million years.

EXTRA
THE CURSE OF FAT BROUGHT ON BY THE BIG BRAIN

Humanity has always endeavored to secure a lot of calories to maintain our big brains. Even with all that effort, humans have sometimes struggled to have enough food. Some countries, however, have more than enough food to sustain their populations (at least right now). What's more, even though many of us in those fortunate countries know it is bad for our health to overeat, we cannot help but be drawn to consume the high-calorie foods in front of us.

The documentary *Super Size Me* (2004) features Morgan Spurlock, who vows to eat only McDonald's food for a month. When the month is up, he discovers that the toxins in his liver have risen to a dangerously high level, along with many other alarming health indicators. The purpose of this documentary was to show the harm inflicted from eating the food from McDonald's; I, however, was impressed that humans can eat so much harmful food for such a long time. Consider, too, the popular eating contests that take place all over the world, or all the eating-challenge podcasts and shows on the internet.

All animals have greed for food to the point that they may harm themselves or cause an untimely death; interestingly, humans are not at risk of immediate fatality after eating too much fatty food. Instead, humans suffer from more long-term, chronic diseases, such as heart disease and diabetes. Our drive for high-calorie foods served us well in growing and maintaining our brains, but now we're stuck with that addiction, even though we don't need all the food we have access to.

CHAPTER 17

You Are a Neanderthal!

"You're such a Neanderthal!"

If someone said this to you, how would you feel? If you remember what you learned about Neanderthals in school, you'd probably feel insulted. Neanderthals are our relatives who lived in Europe from approximately 300,000 years ago until 20,000–30,000 years ago.* Thus, someone who calls you a Neanderthal is basically calling you a primitive human, or even almost an animal. If Neanderthals are also human and our closest relatives, how did their name come to be a derogatory term?

* There is no single correct answer about when and where Neanderthals lived. The uncertainty comes from there being no complete consensus about which fossils or which sites to call "Neanderthal." Historically, the term referred to fossil hominins that lived in Europe during the last glacial period there (Würm), from 100,000 years ago until 30,000 years ago. But now, there are fossils before and after that time period that some call Neanderthals; furthermore, with DNA data the debate continues about the precise dates of Neanderthal tenure.

Embarrassing relatives

The first Neanderthal fossil was discovered in 1856, before the publication of Charles Darwin's famous book *On the Origin of Species* (1859). Subsequent discoveries of Neanderthal specimens drew attention because of their odd appearance: they looked strange. Soon these specimens were the center of a fiery debate about whether they were related to modern humans or to some more distant ancestor instead, without having a direct relationship to modern humans.

The debate about the relationship between Neanderthals and modern humans continued to receive passionate attention from the field of paleoanthropology well toward the end of the twentieth century. The University of Michigan, where I was a graduate student, was one of the hubs of this debate. During my time there in the 1990s, there were two competing positions: one saying that Neanderthals were related to modern humans, the other saying there was no direct relationship between the two. In the field of paleoanthropology, the prevalent opinion was that Neanderthals were directly ancestral to modern humans. Our knowledge of Neanderthals came mostly through research on fossils, and many noted that several morphological traits observed in Neanderthals were also observed in modern humans, such as a projecting midface or an occipital bun.

On the opposite side of the debate, a minority in the field of paleoanthropology argued that there was no relationship between Neanderthals and modern humans. Yet, intriguingly, the majority of people in general society believed the minority view that Neanderthals had nothing to do with modern humans. At first, I could not understand this. Why hold this belief in the face of data supporting the direct relationship between Neanderthals and modern humans? To many people, though, this was not a matter of data; it was a matter of pride. For many modern humans, Neanderthals were embarrassing relatives. People recoiled at the idea that Neanderthals were in our blood, whether 30,000 years ago or 100,000 years ago. Why?

What triggered the negative emotional response to Neanderthals was a fossil discovered in the early twentieth century at La Chapelle-aux-Saints, France. This specimen had a skull, trunk, arm bones, and leg bones, and was estimated to have a stooping posture that was due to old age and a physically difficult life. But many people instead interpreted this stooped posture as a sign that Neanderthals were primitive and stupid. An article in a London newspaper published in 1909, the year after the discovery, shows a reconstructed picture of the La Chapelle Neanderthal, reflecting the stigmatized impression of a stooped being with hair covering the whole body, mouth half open, and dull eyes set back in overhanging brow ridges. And the forehead was pictured as narrow, flat, and receding.

Neanderthal and the colonial "primitive man"

Wait, the description of the La Chapelle Neanderthal seems somehow familiar, doesn't it? It is the same denigrating portrayal that colonial Europeans first used to characterize the indigenous populations they encountered through imperial expansion. It is the look of "barbarians." This "primitive man" look was already part of the narrative that Europeans had constructed about themselves and others. To Europeans, the people of the colonies were primitive and in need of guidance from proselytizing Christians to become civilized. Colonizing them was an opportunity to reform the uncivilized.

Now let's consider the Neanderthals. Europeans imagined Neanderthals as having hunted with primitive weapons, howling like animals and crouching on the floor of their caves, almost like beasts rather than humans. Neanderthals met and were defeated by the "Cro-Magnons," who had high foreheads, strong chins, and firm mouths, and perhaps looked handsome, much like Europeans imagined that they themselves looked. To the Europeans, Cro-Magnons were truly human, equipped with sophisticated hunting skills, language, and culture. In contrast, Neanderthals were only almost, but not quite, human.

There was a definite similarity between Neanderthals and the colonial peoples that the Europeans encountered. In one case (the Neanderthals), primitive people went extinct at the hands of modern humans because of their primitiveness. In the other case (colonial peoples), primitive people were afforded an opportunity to become civilized after being colonized. The way Europeans viewed Neanderthals thus converged with the way they looked at the people from the colonies. "You are a Neanderthal!" is a derogatory statement precisely for this reason.

The negative image of Neanderthals persisted for quite some time. In the 1990s, research in genetics seemed to support this image. Study after study using crude DNA sampling of modern humans showed that Neanderthals had nothing to do with modern humans. The approach was an attempt to make inferences about our past by examining the DNA of contemporary humans.

Then, a research team at the Max Planck Institute in Germany led by Svante Pääbo introduced a new method of analyzing ancient DNA that had been extracted directly from Neanderthal fossils. Research by Pääbo's team showed that the DNA of Neanderthals and of modern humans did not overlap at all, indicating no admixture between the two populations. This finding meant that Neanderthals could not be direct ancestors of modern humans. And although full genome sequencing of the nuclear DNA was still outside the realm of possibility, the results were the same whether Pääbo's team used 340 bases from mitochondrial DNA, or more than 16,000 bases making up the full genome of the mitochondrial DNA, or 1 million bases from nuclear DNA. In contrast with the old way of studying fossil morphology, the new way of extracting and analyzing ancient DNA gave off the image of modernity and groundbreaking innovation. It also stimulated our imagination, hinting at a future in which the movie *Jurassic Park* (1993) was closer to reality than we thought possible.

As a result, it became a mainstream belief that Neanderthals were not related to modern humans. By the year 2000, it had also become a mainstream opinion that Neanderthals were driven to extinction by modern humans. Indeed, several hypotheses were proposed to explain the disappearance of the Neanderthals at the hands of modern humans. The range of hypotheses varied from direct encounters that turned violent—resulting in carnage of the Neanderthals by modern humans equipped with powerful weaponry—to indirect competition for resources, which modern humans won with their advanced adaptation and reproductive fitness. In any case, one fact stayed the same: there was no admixture, no interbreeding between these two populations.

We speak Neanderthal?

Yet another shocking reversal took place ten years later, in 2010. Pääbo used even more advanced technology of genome sequencing to extract and analyze ancient DNA, this time comparing the genome of Neanderthal nuclear DNA, more than 3 billion base pairs, to a human genome. The results were shocking. The analysis showed that Neanderthals had left a legacy of DNA in modern humans; Europeans have inherited on average 4 percent of their genes from Neanderthals. Europeans were the descendants of Neanderthals, their very bloodline!

More surprising were the kinds of genes constituting that 4 percent Neanderthal ancestry. They are not random, useless genes, but relevant to critical functions of everyday life. For example, among the genes we inherited from Neanderthals are those that control olfaction, vision, cell division, sperm health, the immune system, and muscle contraction. Particularly surprising was the gene FOXP2, which is related to language. Mutation of this gene causes a loss in language ability. The speech abilities of Neanderthals had always been in question prior to Pääbo's groundbreaking research. There had been a long debate about whether Neanderthals could speak, and if they could, to what degree. Could they talk like modern humans, or did they mumble, speaking in a limited way?

Those who had argued that Neanderthals did not possess speech capacity predicted that Neanderthals' FOXP2 gene would be different from the version found in modern humans. When the Neanderthal genome was published, FOXP2 was one of the first things they looked for; amazingly, Neanderthals had the same version of the FOXP2 gene that modern humans have. Did Neanderthals really speak like we do? (Or more accurately, do we speak like Neanderthals?)

Researchers looked for other clues besides genes to explore the level of speech that Neanderthals had. One interesting characteristic of the modern human brain is lateral asymmetry—that is, differences between the left and the right sides of the brain. Several different parts of the modern brain are related to language, particularly a couple parts on the left side. Lateral asymmetry in the brain indicates that one part of the body is used more than the other. And the brain's particular asymmetry tells us whether a person is right-handed or left-handed. If we had data to show that Neanderthals exhibited handedness, we could infer that their brains also had lateral asymmetry—in other words, that the Neanderthal brain was capable of language.

David Frayer, an anthropologist at the University of Kansas, led a team of researchers who approached this problem in an innovative way. They focused on Neanderthal teeth. Neanderthals are well known for using their teeth as tools, beyond chewing food. Their occlusal surface (where upper teeth and lower teeth meet) is often strangely misaligned. If the teeth were used only for chewing food, the occlusal surface would be aligned because it would be made from the upper teeth meeting the lower teeth. A misaligned occlusal surface indicated that the teeth were used in tasks besides chewing food. For example, when cutting meat or tough plants, Neanderthals held tight onto one side with their teeth, but the other side with one hand. Then with the other hand holding a stone tool, they would strike down and cut the material held between the teeth and the hand. Sometimes the angle of the stone tool would be a little off, and the sharp edge would scratch against the teeth surface. And the angle of the scratch would differ between a right-handed tool user and a left-handed tool user.

Examining the angle of the scratch would therefore give us an idea about the handedness of the tool user. An ingenious idea, isn't it? Results of the analysis showed that Neanderthals were predominantly right-handed, in a 9:1 ratio. This ratio is similar to the one found in modern humans, making it more likely that Neanderthals were capable of speech.

Neanderthal within you, Southeast Asian within me

Since then, further research has shown that Neanderthals were not primitive people howling like animals. Neanderthals used tools and survived in an extremely difficult environment. They knew to decorate their bodies using a red dye called ocher, and they buried their dead with care. It is highly likely that they spoke language fluently, like modern humans. Furthermore, cave art, which used to be thought of as a unique invention of modern humans, was likely started by Neanderthals.

Some say that history advances in spirals. Germany, the site where the first Neanderthal fossil was discovered, was home to one of the most racist regimes in the twentieth century. Yet now there is a movement among Germans to acknowledge and celebrate Neanderthals as their ancestors. One can see German youth wearing T-shirts that say, "Ich bin ein Neandertaler," a play on John F. Kennedy's famous quote "Ich bin ein Berliner," from a speech he gave when visiting Berlin in 1963. Does this mean that Neanderthals are being welcomed into our ancestry? I am cautiously optimistic that the racist imagery of Neanderthals is slowly becoming a thing of the past. This racist view, like the racist colonial view of indigenous peoples, is being reexamined and hopefully disappearing. Our societies are finally embracing and celebrating their diversity.

As I continue to think about Neanderthals, I also think about Korea, the country of my birth. Korea is a nation extremely interested in its ancestry, and Koreans tend to validate themselves by depicting their ancestors in heroic scenes. But that's a little strange when you think about it. What makes an ancestor heroic or not?

In school I learned that the ancestors of Koreans came from northeastern Asia, or Siberia. We Koreans generally like the idea that we descended from people at the northeastern end of the Asian continent. But what if we were told that our ancestors came from Southeast Asia? Would Koreans recoil at this thought of having a dark-skinned, smaller-bodied ancestry? And would that feeling of resistance come from the racist bias that many Koreans hold toward Southeast Asians right now? How would such an attitude be different from the bias that Europeans had against the Neanderthals in the beginning of twentieth century? As scientifically grounded as we may imagine ourselves to be, it is also important to remember that our social ideas about ourselves bias the way we pursue questions about our past and our ancestors.

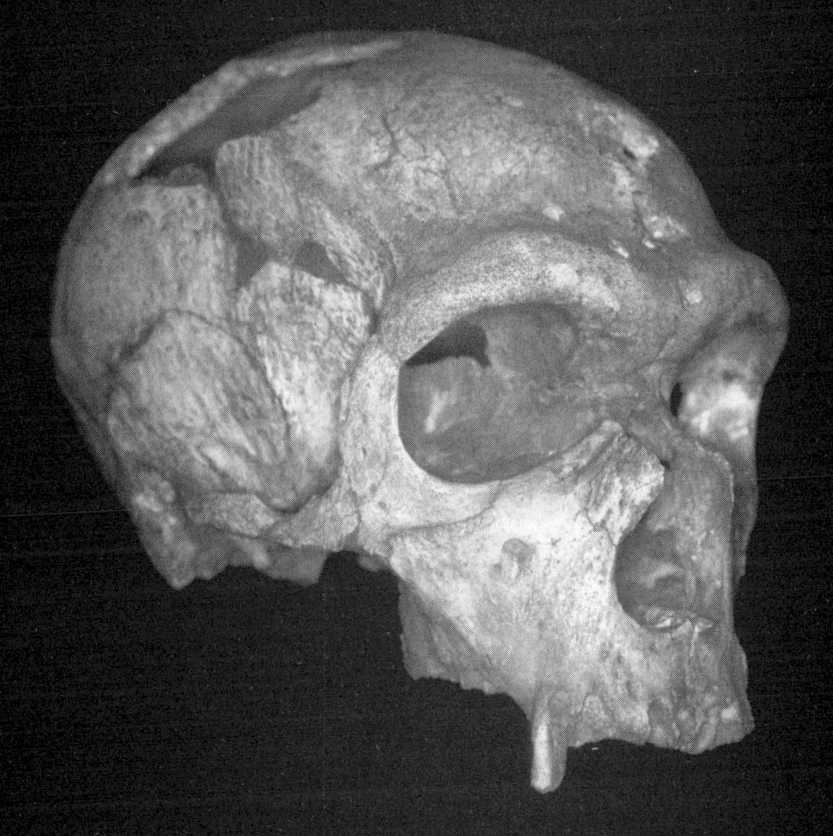

Neanderthal fossil skull discovered in La Chapelle-aux-Saints, France.
(© Milford Wolpoff)

CHAPTER 18

The Molecular Clock Does Not Keep Time

With their big shoulders and thick chest, arms, and legs, Neanderthals appear to have been quite robust. In today's language, you might say that Neanderthals were "buff." Or, if you want to put a more negative spin on it, they were "savage."

Neanderthals share a lot of morphological similarities with modern humans. Although paleoanthropologists now know that Neanderthals were ancestors to modern humans, at one time they wondered how Neanderthals, with such muscular, "savage" bodies, could be directly linked to modern, "civilized" humans.

The idea that Neanderthals were ancestral to modern humans was famously challenged in a 1987 paper published in the journal *Nature* by the research team led by Rebecca Cann, then at UC Berkeley (and now at the University of Hawai'i, Manoa). An analysis of mitochondrial DNA of modern humans from worldwide samples led to the conclusion that Neanderthals and modern humans did not interbreed. As described in the previous chapter, however, Svante Pääbo later showed that humans and Neanderthals did indeed interbreed.

At the time of Cann and her team's publication in 1987 it was thought to be an amazing accomplishment to be able to study extinct ancestral humans using genetics from modern humans, not fossils. Their work was considered not just technologically impressive, but also an incredibly imaginative feat of research design. How could modern-human genes be used to study our relationship with Neanderthals? Cann's team used a method based on the concept of a molecular clock; it involved tracking the time that had passed since the "time of origin" of a gene by counting its mutations.

The geneticists of the time came to be sure that Neanderthals could not be ancestral to modern humans simply because of the timing calculated from the molecular clock. The time of birth of modern humans was calculated to be less than 200,000 years ago. The place had to be Africa: Cann's research showed that Africans were the most diverse in terms of genetics, meaning they had been around for the longest period of time. Since Neanderthals had already been living in Europe since approximately 250,000–300,000 years ago, they could not have been our ancestors. Case closed.

Cann's research triggered a big wave of human evolutionary studies that also used modern-human genes to study the past. Genes from modern humans became a time machine that carried us into the past. In this chapter I want to talk about this genetic method of time travel, which is widely used in anthropology and other biological sciences. You'll first need a little background in evolutionary science and the life sciences, but then you will learn about one of the most important and controversial fields in modern genetics. So please be patient and read on.

Birth of a molecular clock

Deoxyribonucleic acid, or DNA, transmits genetic information in all life-forms. The genetic information is coded in the arrangement of four bases: adenine (A), guanine (G), cytosine (C), and thymine

(T). These four bases are the building blocks arranged within DNA to record and transmit meaning. The process of organizing these bases is like arranging consonants and vowels in a single line, then reading it to see whether it makes sense.

One could think metaphorically of DNA as a long necklace made up of four different colored beads, which are the bases. Within this chain, only a portion can be called a gene, which in turn is made up of units of triplets called "codons." Each codon codes an amino acid. The length of a gene can vary. In humans, a functional gene can be as small as 200 base pairs (since every base is paired with another base, it is customary to say "base pairs" instead of "bases") or as large as 2 million base pairs. Genes are then read by special enzymes and translated into proteins, which play various important roles in the body.

DNA goes through almost an infinite number of replications. We start from a single cell and grow to have almost 37.2 trillion cells by one estimation (some estimate 70 trillion cells). Each cell has its own life span and has to be replaced when it dies or is damaged; an adult human male is estimated to lose 96 million cells every minute.

Every time a new cell (including an egg or a sperm) is made, DNA is replicated. In the process of these replications, errors in copying are inevitable. Life-forms have an amazing ability to correct and prevent errors in DNA replication, but errors still happen. A base can be inserted or deleted within the chain. Sometimes, a wrong base takes the place of a correct one. When this happens, a different codon may result, possibly causing the wrong amino acid to be made, or producing an entirely nonsensical string of bases instead.

Such errors are called mutations. What happens to an organism with a mutation in its genes? The Marvel Comics X-Men suggest that mutations may give certain humans superpowers for fighting crime. Motoo Kimura, a geneticist from Japan, offered an alternative view: in population genetics, we can recognize mutations only if they have no effect on our lives; we cannot recognize mutations that have an impact on a single individual. This is the central statement of the neutral theory, one of the most important theories in modern population genetics. It all sounds rather confusing.

Let me explain. We align genetic sequences (the chains of bases in DNA) from two or more individuals and compare each base on the same location in the genetic sequences of different people, to see whether they all have the same base at that location. Suppose one individual's genetic sequence has the base G (guanine) at the fifth base location, and another individual has a C (cytosine) at the same location. Then we can conclude that a mutation occurred at the fifth base location. We do not know whether C became G, or G became C. We just know that a mutation occurred at that spot. This mutation could have a variety of possible effects. If the mutation occurred on a coding region, it could cause a change in the amino acid and change protein structure. The different protein structure could then either harm or benefit the individual.

If a mutation is harmful to the individual's life and its reproductive fitness, that individual will not be able to leave any offspring, so that particular harmful mutation will disappear from the gene pool sooner or later. When comparing the genetic sequences of living individuals at some later time, we would not notice that a mutation occurred in the past and subsequently disappeared.

If a mutation is beneficial, an individual carrying that mutation will leave many offspring. After several generations, the beneficial mutation will spread, until everyone in the gene pool has this mutation. Once everyone has the same mutation version of the gene, we can no longer discern that a mutation occurred and spread to everyone. Because we recognize a mutation from the *differences* we spot when comparing genetic sequences, we cannot recognize a mutation when the genetic sequences are exactly the same. For this reason, whether the mutation is beneficial or harmful, at some point we will no longer be able to recognize and identify it as a mutation.

Alternatively, the mutation may change the codon but the changed codon may still code for the same amino acid; in this case, the mutation leads to no change in the protein structure and has no effect on the individual's life. Or, the mutation may occur on a noncoding region of DNA and have no effect on the individual's life. These mutations that have no effect on the individual's life stay in the genetic sequence and can be recognized as mutations. They don't disappear from the gene pool because they are harmful, nor do they spread to the whole gene pool because they are beneficial.

The frequency of such a mutation in a gene pool does change over time, however. As time goes by, the mutation either increases or decreases in its frequency, by random process. This means that if we know the frequency pattern of a particular mutation, we can determine how long the mutation has been around and, by extension, the population that carries it. The neutral theory, based on this logic, made a significant contribution to the field of population genetics in the twentieth century. In fact, population genetics after the 1960s can be said to have flourished on the foundation of the neutral theory.

Therein lies the irony of the neutral theory. Recall that Charles Darwin was the first to propose natural selection as the core mechanism of evolution. In modern biology, evolution is defined as a change in gene frequency over generations. How ironic that a change in the frequency in DNA has less to do with a selective advantage, but more to do with time and randomness!

The number of mutations determines how much genetic diversity a population has. In the 1990s, geneticists made an interesting discovery about the genetic diversity of human populations: it turned out that humans are not very genetically diverse. The 1987 study by Cann and colleagues had concluded that the diversity in human mitochondrial DNA is surprisingly low. And a groundbreaking 1991 paper on population genetics by Wen-Hsiung Li and Lori Sadler, geneticists then affiliated with the University of Texas, was titled "Low Nucleotide Diversity in Man."

Population geneticists interpreted these findings in the framework of the neutral theory. The low level of genetic diversity found in humans was attributed to recent origin of the lineage. In other words, because humans originated recently, there simply hadn't been enough time for many mutations to accumulate. If the human species did originate recently, however, all early-hominin fossil species that had lived all over the world before modern humans must have gone extinct as genetic side branches instead of becoming our direct ancestors. For example, Neanderthals, who appeared in the beginning of this chapter, were once thought to represent such an extinct side branch. Because Asia and Africa did not have many hominin fossils that appeared right before or overlapped with the earliest modern humans, the debate about the origins of modern humans concentrated on the relationship between Neanderthals and modern Europeans.

Another exciting fact came out of this population genetics research: when Africans, Asians, and Europeans were compared for their genetic diversity, Africans showed the highest diversity. Since higher diversity means an earlier divergence of the lineage, it appeared that all modern humans must have migrated out of Africa recently. The theory that modern humans originated in Africa about 200,000 years ago gained a strong body of supporters and, aided by advances in modern biology, became the mainstream idea of human origins. It came to be known as the "complete replacement model."

The molecular clock breaks

Early efforts to study human evolution using genetics all focused on mitochondrial DNA. The reason was that, at the time, it was extremely difficult to deal with nuclear DNA consisting of 3 billion base pairs.

The mitochondrial DNA genome contains more than 16,000 base pairs and was studied in many other organisms as well. Most important, researchers thought mutations in mitochondria had no effect on the individual because mitochondria were outside the cell nucleus. In other words, mitochondrial DNA was deemed "neutral."

In the late 1990s, however, molecular evolution based on the neutral theory started to be challenged bit by bit. It all began with the widely studied mitochondria. Mitochondrial DNA is maternally inherited; if a woman has only sons and no daughter, her mitochondrial lineage will disappear. And so, then, will the mutations that occurred in her mitochondrial lineage. Thus, the actual number of all mutations that ever happened in the past may be much larger than the number of mutations we can observe through mitochondrial DNA of living people. We would be greatly underestimating the time of origin if we used only mitochondrial DNA evidence to reach a conclusion.

Another challenge arose to the supposed predictability of mutation periods in DNA, calling into question timing estimates based on those mutations. For example, suppose a mutation happens every 100 years. If we observe five mutations, we can estimate that 500 years have passed. This is basically how we estimated the time of our genetic origin until now. But what if a mutation happened every 50 years, instead of every 100 years? Then, 250 years would have passed, not 500 years as estimated. Alternatively, if a mutation happens every 200 years, then 1,000 years would have passed. And what if mutations do not happen regularly, as predicted? Then we would not be able to tell time at all. And that is what researchers found. As more research showed that the mutation rate in mitochondria might not be constant, it became more uncertain to estimate the time of origin. The much-prized molecular clock did not keep time.

Another criticism against the neutral theory came with advances in modern biological sciences. The neutral theory relies on the assumption that mutations in the noncoding region of DNA have no effect on the individual. Research started to show that such mutations do, in fact, affect the life and the reproductive fitness of an individual.

The noncoding region had been called "junk DNA" because it was assumed to be meaningless. But more research is showing that junk DNA plays an important role by regulating or signaling other genes. Even the mutations that do not occur in the coding region can have a beneficial or a harmful effect on life. This means that such mutations are subject to selection, and likewise that the molecular clock based on the neutral theory will inevitably be inaccurate.

And that's not all. Since the early 2000s, mitochondrial DNA, previously thought not to affect an individual's life (because it was located outside of the cell nucleus), has been known have an impact as well. Though located outside of the cell nucleus, mitochondria are in charge of cell metabolism. I'll never forget the sense of wonder and excitement I felt as one study after another showed natural selection playing a large role in shaping the genetics of mitochondrial DNA. In hindsight, it now appears ridiculous that we once thought mitochondria, the energy factory of the cell, could be neutral. But back then, such was the power of a prevailing theory.

Mystery of human origins, round 2

Today, research on human origins is in a new phase. All the publications that traced ancestry through mitochondrial DNA may have to be questioned and reexamined. Furthermore, geneticists are no longer limited to tracing time through DNA from living people, but are able to directly extract DNA from fossils and analyze it. In 1997, the first such research done at the Max Planck Institute in Germany completely fit the mainstream idea among geneticists at the time, which was based on the neutral theory: it supported the complete replacement model.

The mitochondrial DNA from a Neanderthal fossil was quite different from that of modern humans. And the analysis based on nuclear DNA, published in 2006, also showed a big difference in genetics between Neanderthals and modern humans. These studies, however, analyzed only a portion of the whole genetic sequence. From 2010 on, the whole Neanderthal genome was decoded, and analyses based on the genome overturned the results from previous studies. New whole-genome studies showed that Neanderthals interbred with modern humans: on average, 4 percent of the genes of Europeans come from Neanderthals. Support for the complete replacement model, based on the neutral theory, began to weaken.

In 2013, it was reported that ancient DNA had been successfully extracted from a 700,000-year-old horse fossil. It seems like it's only a matter of time before ancient DNA from a similarly old hominin fossil can be extracted as well. We are now, amazingly, able to travel through time via our own genes and the genes of ancestral humans.

EXTRA
NO JUNK

Just like the mistaken idea that we only use 10 percent of our brains, the theory of "junk DNA" has been disproved, though for a while it seemed to have clear evidence on its side. The human genome is made of 3 billion base pairs. When it was decoded in 2001, the shocking discovery was that this unimaginably large number of base pairs contained only about 20,000 functional genes, genes that code for protein synthesis. That is, only about 1 percent of the 3 billion base pairs appeared to be functional, with the remaining 99 percent doing nothing. The small proportion of genes in the genome seemed to support the concept of junk DNA, which had gained a lot of traction not only among geneticists, but among the general public as well.

What was this 99 percent doing just lying around? Cells multiply and divide numerous times over our lifetime through mitosis to make new cells for our body (somatic cells). Numerous multiplications and divisions happen through meiosis when we make gametes (sex cells). Each time we copy 3 billion base pairs for mitosis and meiosis, are we copying useless information?

Thanks to ongoing research, we are learning that junk DNA has important functions after all. While it does not code for protein synthesis, it does send signals to start or stop making proteins. When this signaling, controlled by the junk DNA, goes haywire, the result can be runaway cell reproduction and cancer.

Whether DNA or the brain, over time we have mistakenly assumed that certain body parts have no use merely because we do not know *how* they are used. Human knowledge spans vast areas, but there are still arenas in which we know nothing, and we need to keep learning before we judge that those arenas contain nothing worth knowing.

CHAPTER 19

Denisovans: The Asian Neanderthals?

Neanderthals are perhaps the most studied hominin species in the field of paleoanthropology—probably because we have the largest body of data on them and they lived close to us in time. In addition, the majority of paleoanthropologists are of European descent, searching Europe for our ancestors. Only recently were other hominins, possibly in large numbers, found to have lived in Asia and Siberian Russia at about the same time as Neanderthals. They are called the "Denisovans." Denisovans are closely related to modern humans and share a common ancestor with Neanderthals. Thus, Denisovans are the third hominin group to feature prominently in the heated debate about the origins of modern humans, after Neanderthals and modern humans.

The fossil remains of Denisovans were first discovered in a cave called Denisova, near the Altai Mountains of eastern Russia, on the boundary between Russia and Mongolia. Paleoanthropologists have long wondered whether hominins other than Neanderthals, the well-known hominins in Europe, lived on other continents, such as Asia. In the 1970s and 1980s, when I was going to school in Korea, we all learned about the standard stages of human evolution: *Australopithecus*, *Homo erectus*, Neanderthals, and finally *Homo sapiens*, which appeared in the continents of the Old World (Eurasia and Africa), in an orderly sequence. Now we know that *Australopithecus* is found only in Africa, but back then there were excavations to find *Australopithecus* in Asia, with China leading the effort. Even as recently as the 1970s, papers published in China announced the discovery of an *Australopithecus* fossil there, although such claims remain unconfirmed.

Given the desire to find *Australopithecus* in Asia, imagine how much more exciting it would be to find a Neanderthal there! France has been actively involved in excavation efforts in northeastern Asia to find the missing hominin that would fill the gap of time occupied by the Neanderthals in Europe, and so has China. Even into the twenty-first century, papers published in China, North Korea, and Russia refer to *Homo neanderthalensis*, a species designation for Neanderthals, or *Homo sapiens neanderthalensis*, a subspecies designation for Neanderthals. This assessment is often based on traits found in the newly discovered fossil remains that are considered to be associated with Neanderthals, such as projecting eyebrows or an occipital bun.

Tools without toolmakers

Despite these efforts, no fossil found thus far in Asia can clearly be called a Neanderthal. In fact, few fossils of any kind exist in northeastern Asia from the period of the Neanderthals in Europe (about 100,000 years ago until about 30,000 years ago). You could almost call this period the "dark ages of hominin fossils" in northeastern Asia. And we're not just talking about fossils. Not even the archaeological stone tools associated with Neanderthals—Mousterian tools—have been found in Asia during this period.

Until recently, the most northeastern part of Asia where Neanderthal remains have been discovered has been the Caucasus region of western Russia. In the Mezmaiskaya cave, a Neanderthal fossil of a young child was discovered in an archaeological site with an estimated date of 40,000 years old. Because no Neanderthal signature was found anywhere east of this or in Southeast Asia, paleoanthropologists inferred that Neanderthals must not have traveled beyond the Himalayas.

Does this mean that hominins did not live in Asia from 100,000 to 30,000 years ago? Was there a gap after *Homo erectus* disappeared, until the migration of modern humans who left Africa? Many scholars thought so, until 2010, when a new discovery was made—not of a Neanderthal or modern human, but of an unknown third hominin. This new hominin shared a lineage with the European Neanderthals but was also distinct enough genetically to receive its own special designation as the Asian Neanderthal, the "Denisovan."

Actually, the idea of a third hominin had been circulating among some groups of archaeologists for quite some time. In the Russian Altai region, where the Denisova cave is located, there was evidence of hominin occupation (stone tools and ornaments) since 100,000 years ago. From the continuous distribution of archaeological artifacts, it seems hominins did not leave this region, but continued to live there for at least 100,000 years. They left behind several different kinds of stone tools, and intriguingly, the tools changed significantly about 70,000–80,000 years ago. Discovered in Kara-Bom and Ust-Karakol, these tools were blade tools, a characteristic stone tool of the Upper Paleolithic, widely referred to as the "stone tool of modern humans." Actual modern humans, however, do not begin to appear in the fossil record until about 40,000 years ago. There was a mystery here for paleoanthropologists: another hominin must have made these tools.

The period from 50,000 to 30,000 years ago is an interesting time. Hominins of that period hunted during the summer and survived the winter in caves, protected from the harsh cold. Denisova was one of those caves in the Altai Mountains. In the ceiling of the Denisova cave is a natural hole that could even function as a chimney. It was a perfect place to camp out in winter, with a fire. Naturally, this cave was often used by hominins. Strangely, though, hominins during this period left only two cultural signatures, both of them characteristic of modern humans: hunting tools probably used as spearheads were found, along with a necklace made from animal teeth and a bracelet made of stone material. But no hominin fossil was found, and the identity of the makers of these archaeological finds remained a mystery.

Then, in 2008, a tiny fragmentary bone was discovered, smaller than a pea. It looked like a pinky bone, but no one paid much attention to it. Because no other hominin fossil had come from the cave, people didn't think this bone was from a hominin. Perhaps it was came from a cave bear or another animal living in the cave.

Our unknown Asian relatives

In 2010, analysis of ancient DNA extracted from this bone showed that it had belonged to a girl of six or seven years old, with unfused growth plates. Her DNA proved that she was human, but different both from modern humans and from Neanderthals. Neanderthal fossils of a similar date from the Mezmaiskaya cave in Russia and the Vindija cave in Croatia were both also different from the girl's DNA. All this means there was another, Neanderthal-like lineage, distinct from the European lineage.

Unlike nuclear DNA, which is the same in every cell of a given individual, several different mitochondrial DNA lineages can exist within one individual. Three mitochondrial genomes were extracted from the bone found at Denisova. The mitochondrial DNA extracted from the Asian fossil was different from that of any Neanderthal from Europe or the neighboring Altai region. It was now clear that researchers were dealing with another hominin lineage, in addition to modern humans and Neanderthals. Paleoanthropologists named the hominins represented by the bone (and the DNA) "Denisovans."

Later a molar (a wisdom tooth) was discovered that had a morphology slightly different from that of a modern human or a Neanderthal. A pinky bone fragment and a wisdom tooth do not give enough information to say that Denisovans were morphologically different, much less a new species. Our Denisovan ancestors exist only as genes right now; there are no substantial fossil samples to give us an idea of what they looked like. But in this new era, we can study fossilized species even when only a few fossils are available. We may be living through a time of groundbreaking revolutionary change in paleoanthropology.

Geneticists and paleoanthropologists have explored whether modern humans have Denisovan genes, given the history of admixture with Neanderthals. The initial results were strange. Some modern humans did have Denisovan genes, but they were found far from the Denisova cave, all the way in the south among the Melanesians of Papua New Guinea and the Solomon Islands. On average, 4–6 percent of their genes were found to be inherited from Denisovans. They also shared a percentage of their genes with Neanderthals; they are modern humans, but as much as 8 percent of their genes could have come from these archaic humans.

In contrast, research suggested that only weak signals of Denisovan DNA, less than 4 percent, were found in East Asian populations, who are geographically much closer to the Denisova cave than the Melanesians are. Considering that Neanderthal genes are most strongly present in modern Europeans, who live where the Neanderthals used to live, this Denisovan geographic mismatch of DNA is a strange finding.

How can we explain these results? The most convincing hypothesis is this: Denisovans must have been widespread all over the Asian continent in the Late Pleistocene (approximately 125,000 years ago until 12,000 years ago). Then they exchanged genes with modern humans out of Africa (meaning, they interbred), and the Denisovan genes that had an adaptive advantage remained in the DNA of modern humans. The Denisovan genes found in modern humans are most often associated with the immune system. Moreover, updated Denisovan ancestry studies have suggested widespread Denisovan admixture of between 1 and 3 percent in populations from northwestern Russia to northeastern Asia and across the Asian continent, as well as in some European and African populations. Recently, a gene found among Tibetans to which high-altitude adaptation is attributed was found in the Denisovan DNA as well, spurring the idea that whatever the case may be, Eurasians and some African populations appear substantially admixed with Neanderthals and/or Denisovans.

How do we explain these surprising patterns of Denisovan admixture? Could modern Asians have moved into Asia after the humans bearing Denisovan genes migrated to Melanesia and then later spread the Denisovan genes around? No concrete conclusion is possible yet, since the discovery of Denisovans is still quite new. It's also possible that Denisovan genes are, indeed, widespread among modern Asians, but the mechanisms of their dispersal are poorly understood. It's only a matter of time before larger samples and, hopefully, more fossils are discovered. Right now the data are insufficient—all the more reason to look forward to further research about our origins.

Three hominin species, one Denisova cave

Four hominin fossil specimens have been found in the Denisova cave so far—a pinky bone fragment, two wisdom teeth, and a toe bone—three of which have been attributed to Denisovans. Interestingly, genetic analysis of the toe bone showed affinity with Neanderthals:

its DNA looks similar to Neanderthal DNA. In its shape, the toe bone was also most similar to the toe bones from other Neanderthals found in Iraq. In a cave only sixty to ninety miles away from Denisova cave, a Neanderthal fossil and stone tools were also discovered, dated to be 45,000 years old. It looks like Denisova cave provided shelter throughout the ages for Denisovans, modern humans, and Neanderthals.

From all of this archaeological investigation, here's what we know so far: About 70,000–80,000 years ago, Denisovans were living in the Denisovan region of the Russian-Mongolian border. Then, 45,000 years ago, Neanderthals moved into the region (or perhaps they were there already). These Neanderthals left behind their stone tools and a few small fossil remains. Nonetheless, they all appear to have left the area by about 40,000 years ago, and modern humans subsequently took their place. Thus, the Altai region was occupied by three different hominins within a short period of time.

What happened among these different hominins? Did they interbreed and leave offspring behind? Modern-human DNA contains Neanderthal and Denisovan DNA, but it's worth pointing out that Denisovan DNA includes 17 percent Neanderthal DNA. The three hominin lineages seem to have associated in complex ways that we cannot yet fully comprehend.

Research continues to add depth to our knowledge, and with this depth comes complexity. The origin of modern humans was never thought of as simple, but our roots are becoming more complicated and tangled than we ever thought before.

CHAPTER 20

Hobbits

Gigantopithecus, with an estimated body size of over 1,000 pounds, was a giant (see Chapter 13). At the other extreme is a hominin of very small size: *Homo floresiensis.*

There is an interesting legend about the island of Flores, Indonesia. The legend talks about Ebu Gogo, a little person barely 3 feet tall, with large feet and fur covering its body. This legend may have inspired the humanlike hobbit from *The Lord of the Rings,* which has big feet covered with hair.

In 2003, Australian paleoanthropologist Michael Morwood discovered a small hominin fossil on Flores that looks just like the hobbit described in J. R. R. Tolkien's books. The fossil had a very small body and an unbelievably small brain. In fact, the brain was almost smaller than that of a human newborn. Morwood concluded that this was a new hominin species never seen before, and he named it *Homo floresiensis,* meaning "human from the island of Flores." The news media immediately nicknamed it the "hobbit."

The mystery of the hobbit

This was not the first time that a hominin presence had been discovered on Flores Island. Though actual hominin fossils were not found, other archaeological evidence had been discovered throughout years of excavation, starting as early as the 1950s. The archaeological evidence dated as far back as 700,000 years ago, and anthropologists estimated that hominins had lived on the island for at least a million years. Morwood's discovery in 2003 provided fossil confirmation of a hominin presence dating from the period between 60,000 and 18,000 years ago. Compared to their presence on other islands of Indonesia, such as Java, which has hominin evidence from as long ago as 1.8 million years, hominins appear to have arrived at Flores a little late.

A complicated problem, however, lurks beyond the simple numerical dates. Any map of Southeast Asia shows numerous islands in the South Pacific. These islands are surrounded by seawater, some shallow and some deep, and a deep water line called the Wallace Line divides this area into two regions: Southeast Asia and Australia. North of the Wallace Line, where the island of Java is, the water is shallow, and during the cyclical glacial ages the sea level was even lower than it is now, meaning that most of the islands in this northern region were connected to the Asian mainland. Animals, including hominins, could easily walk to these islands.

The island of Flores, however, is located south of the Wallace Line and is surrounded by deep sea. Even when the sea level was lowest during the glacial ages, Flores was still an island surrounded by deep water. Flores Island was separated from the Asian continent even when other islands in Southeast Asia were connected, and it was therefore accessible only by boat. Given this situation, it was a big deal that hominins had been living on Flores Island for the last million years. How did they get there? No wonder anthropologists were interested!

It is not known how hominins arrived at Flores island, or whether they did so intentionally or by accident. Either way, once they arrived, it would have been very difficult to leave the island.

The time period of Flores hominins is similar to the period during which hominins started to live in Australia. Like Flores, Australia is located south of the Wallace Line, and it is isolated in the middle of the ocean. Modern humans, now known as Australian Aborigines, crossed the ocean and settled in Australia between 60,000 and 40,000 years ago. This raises a series of questions. Are the Flores hominins modern humans who crossed the ocean early, like the Australian Aborigines, or are they an extinct relative? The fact that there was only one complete skull among other partial skeletal remains on Flores only enhanced the mystery of the hobbits. Ultimately, scientists were divided into two camps on the issue of the Flores hominins. One side argued that while *Homo floresiensis* looked somewhat strange and perhaps even had suffered from some disease, it was nevertheless a modern human; the other side argued that *Homo floresiensis* was not a modern human, but a new species with a very small body and brain.

The debate centered on the skull's size and shape. A comparison of brain size led to the conclusion that *Homo floresiensis* was unlikely to be a modern human. The brain of the Flores hominins was slightly more than 400 cubic centimeters (cc) in size, smaller than that of a human newborn or an adult chimpanzee. Even modern humans with dwarfism do not have such a small cranial capacity. Individuals with dwarfism are often barely 3 feet tall, but their brain size is no different from that of an average modern human. The Flores hominin could not be considered a simple example of a modern human with dwarfism.

Some researchers suspected microcephaly, a condition, recently associated with the Zika virus, that leads to a small brain. To confirm, they looked for symptoms of microcephaly that are expressed in the body through brain morphology and retarded bone development. But the results were unclear. The proponents of the Flores hominin as a new species argued that the Flores brain and a microcephalic brain were different, drawing from research published in 2005 by anthropologist Dean Falk and her research team at Florida State University. Using micro-CT technology to examine the inside of the Flores skull, they concluded that although the Flores brain is as small as a microcephalic brain, the morphology is quite different.

Columbia University anthropologist Ralph Holloway countered that the Flores skull shape was a result of deformation after being buried underground, and the debate returned to its starting point. Interestingly, these two scholars had been rivals in the 1980s debate on Taung Child, the *Australopithecus africanus* specimen that I mentioned in Chapter 11. Thirty years later, in 2010, they were locking horns again, this time over the Flores fossil.

A trick up its sleeve: new evidence from wrist bones

Since the study of the skull was inconclusive, anthropologists focused on other traits. One of those was whether Flores hominins made tools. Many did not agree that anyone with a brain smaller than that of a newborn or a chimpanzee could make stone tools. The stone tools found on Flores are similar to the Oldowan tools from 2 million years ago. Some argued that there was no way anyone with a 400-cc brain could make tools like that.

Other scholars focused on the overall body size. The leg bone of the Flores hominin has a length similar to the leg bone of Lucy, the famous *Australopithecus afarensis* specimen. But it is also similar to that of some of the smallest peoples among modern humans, such as the Aka of Africa, or the people of the Andaman Islands. This observation led anthropologists to argue that Flores was a miniature version of modern humans, smaller because of developmental problems. To prove their point, paleopathologists noted that the Flores arm and the leg bones were thinner toward the sides, and that they were left-right asymmetrical. They also noted that the tibia (calf bone) was curved. All this could indeed be evidence of malnutrition and stunted development. But it was not definitive evidence. The curvature of the tibia is within the normal distribution of modern humans, and the asymmetry of arm and leg bones could have been a result of deformation after death.

The conclusion of the debate was precipitated unexpectedly by a very small bone: the trapezoid bone, one of the smallest wrist bones. Two such bones were discovered. Since the trapezoid bone forms soon after fertilization, it is not subject to any developmental disturbances that might occur beyond three months of gestation. Therefore, its morphology can shed light on whether the individual is a modern human, regardless of nutrition and development.

Results of the analysis showed that the Flores wrist bone was similar to that of the early hominins of the Pliocene, who had just begun to make stone tools. This bone suggested that the Flores hominins were more similar to nonhuman apes than to humans. There are now more data to support the position that the Flores hominin is indeed a new species, different from modern humans.

Why, then, did the Flores hominin have such a small body? One prominent hypothesis suggested that insular dwarfism holds the key. Animals isolated on an island experience an unpredictable set of selective pressures, different from what mainland animals face. For example, elephants become smaller, while rats and Komodo dragons become bigger. Additional research is needed to confirm this hypothesis. For instance, we need to know whether being isolated on the island led to a small body, or if the Flores hominins arrived at the island with a body that was already small.

A descendant of Australopithecus?

The lengths of arm and leg bones of the Flores hominin and the shape of the pelvis are all similar to those of *Australopithecus afarensis* and *Australopithecus africanus* found in Africa. The brain and body size also resemble those of *Australopithecus.* These facts were quite puzzling to paleoanthropologists. If a fossil specimen that looked like Flores had been discovered in East Africa in a 3-million-year-old deposit, like Lucy, there would be no problem fitting it within what we already knew of early hominin evolution. But the place of discovery was South Pacific Asia, and the Flores hominins lived at the same time as modern humans, whose brains are 1,400 cc on average. How is this possible?

Another set of fossils presented a similar conundrum. They were discovered in Dmanisi, Georgia, between 1991 and 2005. The Dmanisi fossils also were found in a deposit with a date similar to that of *Homo erectus,* but their brain and body sizes were similar to those of hominins far older, *Australopithecus.*

What if the Flores and Dmanisi fossils belonged to the *Australopithecus* genus? That conclusion would indeed be revolutionary, changing what we know of human history completely, because it goes against all mainstream understanding of human evolution and migratory patterns. Our current model of human origins says that from *Australopithecus* in Africa, which had a small brain and a small body, originated a lineage with a bigger brain and bigger body. The new lineage ate meat. Then came the genus *Homo,* which began to travel outside of Africa. *Homo erectus* is famous as the first world traveler, and it was followed by modern humans.

This model implies that *Australopithecus* stayed in Africa because it simply couldn't leave; its brain and body were too small. But if the Dmanisi or Flores hominins are descendants of *Australopithecus* that left Africa, this model breaks down. In particular, it could lead to the conclusion that the genus *Homo* originated in Asia from one of the descendant lineages of *Australopithecus*, lending significant support to the Asian-origin model. Such a shift would cause a big wave of changes in human-evolution studies.

For now, all we have is a tantalizing story: Three million years ago, some populations of *Australopithecus* perhaps left Africa, followed the prairie land, and migrated into Eurasia. Some of them ended up on Flores Island, in Indonesia, via unknown means. This descendant population of *Australopithecus* was isolated on the island, surviving until fairly recently in human development, and was eventually rediscovered as the "Flores" specimen.

For this story to be tested properly as a hypothesis, more evidence is needed. One complete skull is not enough. Michael Morwood, the discoverer of Flores, passed away in 2013. Who will next discover another skull of equal impact?

EXTRA

THEY MEET AGAIN: DEAN FALK AND RALPH HOLLOWAY

Two of the researchers mentioned in this chapter—Dean Falk of Florida State University and Ralph Holloway of Columbia University—have been rivals for decades. Both scholars studied hominin brain evolution using endocasts, meaning casts of the inside of the skull. Their first academic confrontation was in the 1980s.

Holloway argued that Taung Child, *Australopithecus africanus*, had an enlarged occipital lobe, which was indicated by the location of the lunate sulcus in the brain, which he found to be lower in Taung Child than in nonhuman apes. He contended that the lower position of the lunate sulcus

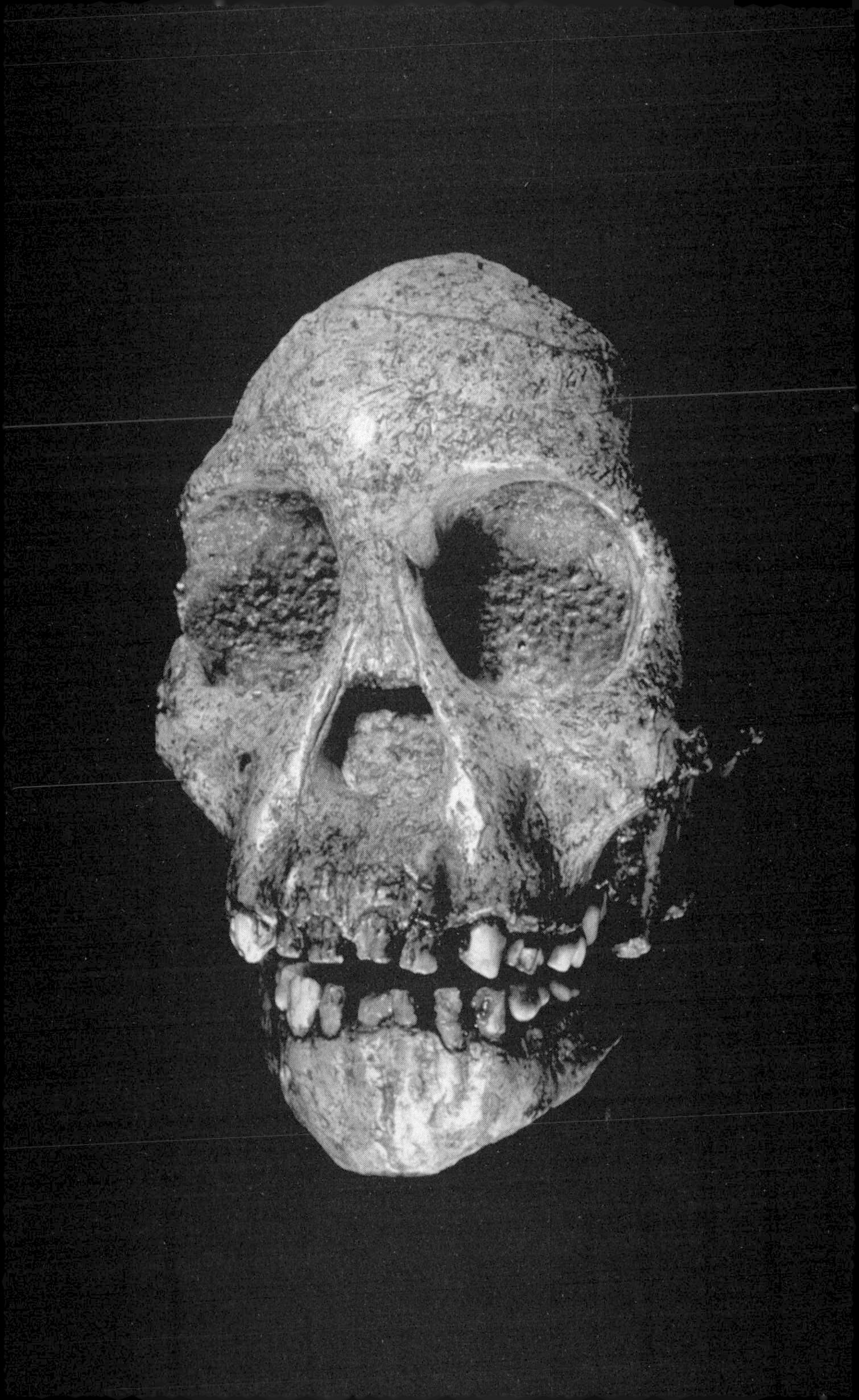

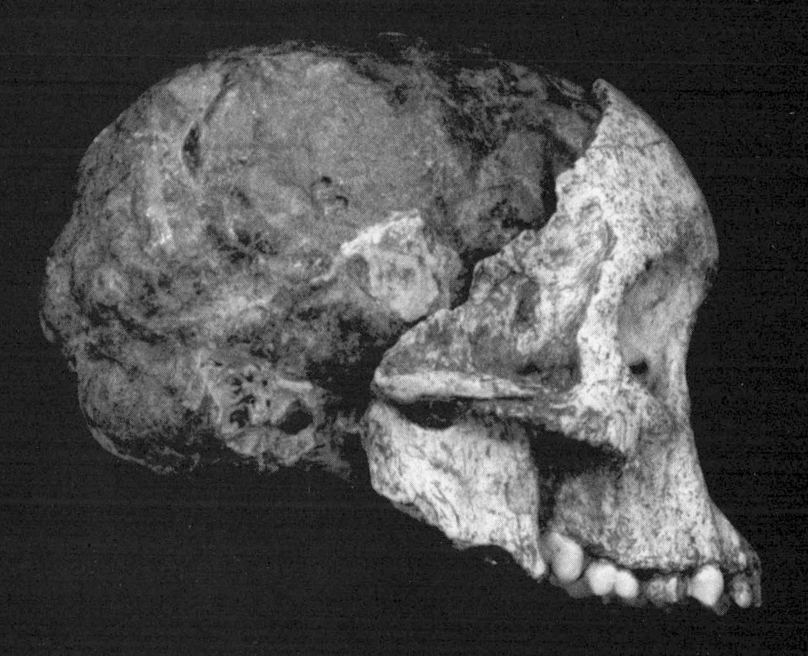

Two views of fossil skull of *Australopithecus africanus*, “Taung Child,” discovered in Taung, South Africa. (© Milford Wolpoff)

was due to the enlarged occipital lobe. The small brain of Taung Child had been cited as evidence against the idea that *Australopithecus africanus* was a direct ancestral species to humans; Holloway claimed that although Taung Child's brain was small, its structure was similar to that of the modern-human brain, and therefore the specimen supported the hypothesis that *Australopithecus africanus* was a direct ancestor of humans. Dean Falk countered that the lunate sulcus was not lower in Taung Child and, furthermore, that its position was not related to the size of the occipital lobe.

This exchange led to a heated debate that continued for twenty years. The dispute was so famous that I remember one day at a conference I was attending as a graduate student, people were whispering because Falk and Holloway had exchanged hellos as they passed each other in the hallway. This was the first time in decades that they had been seen speaking to each other, people commented. The truce did not last very long. Several years later, the two scholars would engage in battle again, this time over the Flores fossil.

CHAPTER 21

Seven Billion Humans, One Single Race?

One of the most sensitive and debated concepts in anthropology is race. Some of you may think that, as a society, we have already decided that all humans in the world belong to a single race and that race is just a racist concept without merit, case closed! But in reality, the case is far from closed. The topic is still hotly debated, and continued research has not helped close the gap between different positions. Our view of the 7 billion humans (and counting) on this Earth is going through a complete overhaul.

It is not clear when and where the concept of race originated. There is a long history among human groups of referring to members of one's own group as "humans" and members of an outside group as nonhuman "barbarians." Evidence of this practice has been found all over the world in written history, and even among populations without written history.

The modern concept of race as it is often used today arose within the last couple of centuries. Europeans explored the world in the fifteenth and sixteenth centuries, "discovered" the New World, and "pioneered" it. By the time Darwin published his book *On the Origin of Species* in 1859, it was widely known in Europe that there were people living in Africa, Southeast Asia, Australia, and the Americas who looked different from Europeans. Europeans debated whether these other people could be considered fully "human." Generally speaking, Europeans settled on the use of three race categorizations: White (Europeans), Black (Africans), and Yellow (Asian), with indigenous people being variously sorted into different racial categories of Malay, Native American, and the like.

In the nineteenth and twentieth centuries, the biological meaning of race was debated. The most extreme position was that a different race constituted a different species, and therefore there were three different species of humans in the world. This view implied that people of races other than Europeans/Whites were not human, and therefore that these different races should not be able to bear children together (conveniently ignoring the many interracial children that resulted from European plantation owners forcing themselves upon their African slaves).

As more Europeans visited more places in the world, the range of variation in different people they encountered increased exponentially. Some people began to wonder whether there were perhaps more than three races. If so, how many more? Five? Seven? At the turn of the twentieth century, the rise of eugenics and the interest in racial purity prompted some European researchers to make it their mission to find out for sure how many races there were in the world.

There are no races, only humans

Researchers' efforts to delineate different groups of distinct races proved, in the end, fruitless. The argument that race is a biological concept like species has no convincing evidence to support it. Extremely long periods of isolation during which different selective pressures challenge different segments of a population need to take place in order for new species to arise. How long does this isolation need to last? Let's consider the case of the Australian Aborigines. Modern humans first arrived in Australia approximately 60,000 years ago. And until the arrival of the Dutch in the seventeenth century, the Aborigines were mostly isolated for 50,000–60,000 years.*

Perhaps this is the reason Australian Aborigines are often described as having a unique appearance. The Europeans who first met the Aborigines questioned their humanness and forbade intermarriages. History shows that neither the law nor the sense of difference prevented intermixing; Australians and Europeans gave birth to many children from their mixed unions. Biologically speaking, however, members from different species cannot have viable offspring.† Moreover, even if offspring result from two different but closely related species, they will not be fertile. A popular example is the infertile mule, born from a union between a horse and a donkey. If Australian Aborigines were truly a different species from Europeans, the children born between them would be either impossible or sterile. And neither of these scenarios is true. If even a 60,000-year period of relative isolation did not lead to a new species, then it is extremely unlikely that a new species would originate from any humans in isolation.

If a race is not like a biological species, researchers keen on proving that races exist as a natural category might then consider subspecies, a subclassification within the same species. A subspecies is a population that has been isolated to the point that it is on a trajectory to become a different species if the isolation continues. Subspecies are sometimes defined as sharing less than 85 percent of their genes with the other subpopulations of their species designation, in order to be considered on their way to becoming a different species.

* There is a big debate about the degree of isolation of the humans in Australia. Probably humans continued to arrive, but it would have been difficult to leave once they were there.

† In light of research on hybridization and hybrid species, this classic definition of species may have to be changed.

Subspeciation is a rather abstract, perhaps vague, concept. The condition for a subspecies, as with a fully different species, is prolonged isolation. It is difficult to assess the level and duration of isolation necessary to define a subspecies, and it is often unclear where a subspecies ends and a new species begins. Furthermore, applying the concept of subspecies to humans is problematic. Humans are never isolated from each other for long; just think of all the broadcasters and photographers who continually search for the most isolated groups of humans and then report on them to the rest of the world. Humans have never been able to continue in a state of isolation long enough to support the concept of subspecies. Humans are a reportedly 99.997 percent genetically similar across all groups, making us exceptionally closely related, compared to other animals.

Finally, there is a simple logistical challenge in clearly diagnosing whether differences in races are at the level of different species. How would you go about proving it? Would you attempt to mix random pairs of humans, just to see if they can interbreed? Clearly this is not ethically feasible, and it would be incredibly costly too. Such manipulation is also morally questionable when it pertains to species closely related to humans. For example, could we try interbreeding a human and a chimpanzee, just to find out whether they belong to different species? In the 1920s, during a particularly dark time in the history of biological sciences, one researcher, Ilya Ivanovich Ivanov, tried to create such a creature, a "humanzee."

Today we can determine species relatedness through morphology and genomic comparison, without resorting to such crude methods. If two populations belong to the same species, they share the same gene pool and consequently will appear similar. We can see from far away whether we're looking at a human or a nonhuman animal. Although everyone looks different, humans share certain traits that differentiate us from other animals, and we know them almost intuitively.

What if two populations become separated and the state of isolation continues for a long period of time, ultimately leading to a separation of the gene pools? As time goes by, members of the two populations will start to look different. Since the populations do not exchange genes, the differences accumulate. As this situation persists, the two populations become different subspecies, and ultimately different species. This is why we can recognize different species by the different appearances between groups. The problem, however, is that there is no way to tell how much difference there should be for a different species. One of the classic questions in paleoanthropology is, "Are these two fossil specimens too different to belong to the same species?" What we perceive to be different—for example, skin color—may or may not be substantial enough to indicate different species.

Currently, nobody thinks different races are different species. Humans do possess varying traits across the globe, but the variation is distributed in such a way that different races cannot be delineated. Some traits match up with geographic distributions. For example, Asians have a higher frequency of shovel-shaped incisors. But other traits are distributed regardless of geographic groupings, or occur along a continuous spectrum such that there cannot be clear divisions. For example, skin color varies in a very gradual way. We cannot draw a clean and clear line that separates "white" skin and "black" skin. To take another example: some people can taste PTC, a chemical with an extremely bitter taste, and others cannot, but the tasters and the nontasters do not align with any traditional racial classifications. Anthropologists have reached a consensus that race is not a biological concept, but a historical, cultural, and social concept.

Neanderthals, Aborigines, and the species question

Applying this view to the debate on the origins of modern humans brings up an interesting question: How different did Neanderthals and the Upper Paleolithic Europeans (those considered to be the first modern humans in Europe) look? Did they look too different to be considered the same species? What if they looked different, but the level of difference was within the range of expectation for a single species? Then there would be no reason to classify them as different species. We can devise ways to measure difference with quantitative rigor, but the question still remains, How much difference should there be for different species? Comparing Neanderthals with modern humans is no exception.

Furthermore, when comparing Neanderthals with modern humans, the first thing we have to do is decide which population of modern humans to look at. And that is not a simple problem. Should the modern humans compared with Neanderthals include representative samples from all major continents—Europe, Asia, Africa, Australia, and the Americas? How could a representative sample be defined?

This question was formally debated some time ago. In the 1980s, two rivals—Christopher Stringer from the British Museum of Natural History and Milford Wolpoff of the University of Michigan—exchanged opinions about the definition of modern humans. Arguing that the debate about the origins of modern humans needs an accurate definition of modern-human morphology, Stringer listed traits that could be used to identify modern humans. In other words, he proposed the conditions of being human. Except, there was a problem: applying those "conditions" excluded a significant portion of modern humans, such as the Australian Aborigines. Were they not humans too? Wolpoff argued against such a list because of its exclusive properties and racist implications. All living human populations today are clearly modern humans. Any list of human traits should include all living humans today, or at least most of them.

What spurred such an absurd exchange? Let's go back to the very idea of the "definition of modern humans." As mentioned earlier, Australian Aborigines have a distinct look. This different appearance is not surprising, considering the long history of isolation, perhaps as long as 60,000 years. If we include Australian Aborigines under the *Homo sapiens* umbrella, should we not include other humans (extinct hominins as well) that also appear different from us?

This question is directly relevant to Neanderthals. Neanderthals also have a distinct look, but not quite so different that you could easily pick a Neanderthal out of a crowd. The appearance of Neanderthals, in other words, is well within the range of variation that we see among humans today. The suggestion, then, is that Neanderthals should be included with modern humans. Furthermore, recent research shows that Neanderthals and modern humans interbred and produced viable offspring. As a consequence, Neanderthal genes are present in human populations all over the world. Is it still correct to divide these two groups into different species? The debate about whether to call Neanderthals *Homo neanderthalensis* or *Homo sapiens neanderthalensis* continues.

Did humans really originate only in Africa?

The debate between Stringer and Wolpoff has cooled down, but it triggered a more fundamental argument—not about who should be included under the "human" umbrella, but about the origin of the human species.

Considering the enormous amount of human variation, I cannot help but wonder whether the complete replacement model, which posits that modern humans originated at one point in time from one place, replacing everyone else, even *could* be correct.

My position is represented by a different model. I think modern humans originated not in one place, but from various places; I think modern humans did not appear as a single population that then spread across the globe, but rather as different populations in different regions that encountered one another as they moved around, and intermixed genetically, evolving as one species. And this process resulted in the vast range of variation that we see today among humans in different places, even though they are all members of the species *Homo sapiens*. This model is called the multiregional evolution model, proposed in 1984 by Milford Wolpoff, Xinzhi Wu at the Institute of Vertebrate Paleontology and Paleoanthropology (IVPP), and the late Alan G. Thorne at the Australian National University. The model's argument—that Neanderthals and modern humans interacted, interbred through gene flow, and continued to evolve as a single species—is also compatible with the recent research in genetics.

Until now, we have talked about our ancestral relatives near and far, and their beginnings and endings. But we do not know even basic information about the rise of *Homo sapiens*. Of all the questions and problems to be solved related to human evolution, perhaps the most interesting and difficult one involves ourselves.

EXTRA
SCIENCE AND POLITICS

In the 1990s, when the debate about the origins of modern humans between the proponents of the complete replacement model and those of the multiregional evolution model was at its peak, the situation ultimately became political—or perhaps it already was. The proponents of each side implicitly attacked the other side as "racist."

Supporters of the complete replacement model argued that modern humans recently originated in Africa, and therefore that the enormous range of variation we see among different people traces its origin to recent human evolution; we are all brothers and sisters under the skin.

Those on the side of the multiregional evolution model argued that human species was not divided into different races but instead has undergone continuous genetic exchange and interbreeding; we have been brothers and sisters for a very long time. They also argued that originating recently in Africa as a new species, and then spreading all over the world and replacing other hominins without intermixing, implies a worldwide bloodbath as humans drove all others to extinction—an argument that plays racist fears against Africans and colonialism against indigenous peoples.

This exchange took place not in published papers, but in seminars and private conversations, off the record. The heated debate, which went beyond empirical support of the data and was influenced by political orientation and sensitivity, reveals that scholars, who have been trained to engage in debates based on cool logic and firm data, are still only human themselves.

CHAPTER 22

Are Humans Still Evolving?

This is a question I often get when I teach. Many people think that humans are done evolving—that since we've accomplished culture and civilization, we are no longer subject to biological evolution. People believe that humans have become so advanced as to transcend our biological dimension.

Leslie White, an anthropologist from the 1960s, wrote, "Culture is the extra-somatic means of adaptation for the human organism." This means that humans adapt to the environment through culture. According to this view, as culture and civilization advance, humans will adapt to the environment through tools rather than their bodies. For example, when it is cold, we turn on central heating rather than developing a thick layer of fat under our skin. As culture continues to advance, the pressures to adapt to the environment through changes to the body are reduced. This makes sense, sort of, but is it true? Have we gone beyond the mechanisms and laws of evolution?

This view reminds me of a conversation I had in the 1990s when I was writing my doctoral dissertation. One of my colleagues, a cultural anthropologist, asked me what my thesis was about. I responded that I was looking at the changes in sexual dimorphism in body size (body size differences between sexes) in human evolutionary history, through fossil data. Her response surprised me a lot: "Difference between sexes? How could you know from looking at the bones, since sex is a social construct?"

This was becoming the mainstream idea in the 1990s, at least in the field of anthropology. People thought that humans were solely cultural beings; nothing was biological. Some even thought that concepts such as the body and genes were also sociocultural concepts, not biological. It seemed as if humans were trying to completely separate themselves from biology.

Culture accelerates evolution

Although it has been more than 2 million years since hominins started to make stone tools, modern civilization started only after agriculture and animal domestication were established, about 10,000 years ago. With these two innovations, humans could produce food (instead of foraging), and productivity increased substantially. The result was a surplus of food that brought with it civilization and social class structures. The rate of culture change appeared to accelerate.

As humans were increasingly changed by culture, biological evolution seemed to take a back seat. Some geneticists have argued that the genetic changes in humans over the last 10,000 years have been neither advantageous nor detrimental in adapting to the environment; therefore, there has been no natural selection. The core concept of Darwinian evolution—the selection of advantageous traits—was brushed aside as insignificant.

New research in the twenty-first century, however, is changing this view. The human genome was sequenced, and the number of sequenced individual genomes is increasing by leaps and bounds. Now we are starting to have enough data to compare different genomes of individual humans. Through this comparison, we can track specific changes that have happened to specific genes. Contrary to the neutral theory, which argues that selection plays a minimal role in genetics, genes that display recent selective changes have been found. Human genes continue to evolve, and surprisingly, the rate of evolution has accelerated with advances in civilization. And the primary factor that has caused these evolutionary changes is culture.

One such example is lighter skin (see Chapter 7). Hominins have spent most of their evolutionary history in equatorial East Africa. Near the equator, ultraviolet radiation is strong, and the mutation that produces a lot of melanin to block it therefore had a selective advantage. According to this hypothesis, this is probably why we developed dark skin.

Some hominins then spread out of Africa and started to live in the midlatitude regions, where UV radiation is weaker. This was the period of the Ice Age, and the short, cloudy days meant even weaker sunlight than what is found in the region today. Skin with a lot of melanin that blocked UV rays was disadvantageous at these latitudes. As I mentioned in Chapter 7, UV radiation is necessary for synthesizing vitamin D. Without sufficient vitamin D, the human body cannot metabolize calcium, and a deficiency in calcium leads to bone deformation, which in turn threatens survival and reproduction. Consequently, a mutation that minimizes or eliminates melanin production was advantageous for the people living in midlatitude regions, resulting in light skin. This is the "vitamin D hypothesis." If this hypothesis is correct, hominins likely developed light skin soon after they left Africa and started to live in midlatitude regions about 2 million years ago. Because skin color is not preserved in fossils, fossilized data cannot tell us when light skin started. The answer comes from genetics.

One of the genes that plays an important role in skin color was discovered in 1999. Today, we know that more than ten genes are associated with pigmentation in the skin. Interestingly, the frequency distribution of those genes is different by continent. The same shade of skin color may have a different combination of skin color genes, depending on the continent where it appears. Light skin in Europeans and light skin in Asians, for example, have resulted from different collections of genes. The light skin in Europeans first appeared 5,000 years ago, which is much later than the time when hominins first left Africa to go north into Europe. Considering that the first migration out of Africa occurred almost 2 million years ago, 5,000 years ago is incredibly recent. The long gap in time between the exodus from Africa and the appearance of light skin implies that the vitamin D hypothesis does not explain everything.

David Reich and his team of researchers at Harvard University, who discovered that light skin appeared recently, propose an alternative hypothesis. They suggest that perhaps after humans started living in the midlatitude regions, they continued their diet of meat and fish, through hunting and foraging. This diet was rich in vitamin D, so there was no need to synthesize vitamin D through skin. And since eliminating melanin was therefore not important, a light-skin mutation was not particularly advantageous.

When agriculture started about 10,000 years ago, however, there was a substantial change in humans' way of life. Instead of fish and meat, grain became the staple diet, and vitamin D was no longer sufficiently supplied through food. As a result, it became advantageous to synthesize vitamin D through ultraviolet radiation. Now the light skin that let the UV radiation through for vitamin D synthesis was more advantageous than the dark skin that blocked the UV rays, leading to the prevalence of white skin. This is an example of a cultural change (agriculture) leading to a biological change (an increase in the frequency of white skin) through natural selection. Culture did not take the place of biology; on the contrary, culture accelerated biological evolution.

In fact, the idea of accelerated human evolution was already circulating among paleoanthropologists who were studying changes in bone morphology in the 1970s. In comparing European skeletons from the Upper Paleolithic and the Mesolithic, David Frayer of the University of Kansas, for example, discovered that the rate of change in limb bone length was faster in the Mesolithic sample, which came after the Upper Paleolithic. This idea still did not become widely accepted at that time, however, because the prevalent hypothesis then was that culture and civilization slowed the evolutionary rate. But now, several studies have found natural selection contributing to recent evolution in humans. There's even a book on the subject: *The 10,000 Year Explosion*, by Gregory Cochran and Henry Harpending, both at the University of Utah, which came out in 2009.

Medical advances accelerate evolution

Compared with the situation in the Pleistocene, over the last 5,000 years humans have gone through evolutionary changes 100 times faster than previous hominins did. Gregory Cochran, Henry Harpending, and John Hawks (then also at the University of Utah, now at the University of Wisconsin) argue that several factors have played a role in this acceleration. First, population increase is a big reason for rapid evolution. This is a refreshing perspective. Ten thousand years ago, with advances in agriculture, human population increased explosively, and so did the number of mutations. With the same mutation rate, an increase in population means an increase in the number of mutations. Since the number of mutations is related to genetic diversity, increases in mutations mean an increase in genetic diversity. Diversity (variation) is the raw material of evolution: a high level of diversity can fuel more rapid evolution.

Genetic exchanges between different populations also enabled evolution. Hominins from early on have continually exchanged genes among different populations. When agriculture was developed 10,000 years ago, however, nation-states were established, and large-scale wars and migrations started to happen. As large groups of humans began to traverse the Eurasian and African continents, the number of exchanges increased explosively, and so did genetic diversity.

The development of modern medicine also played a role in generating diversity. People who would not previously have survived could now live long and transfer their genes to the next generation. For example, an extremely nearsighted person like myself, who may have lived a very short life if born in a Neanderthal or an early agricultural society, can now enjoy a long and productive life.

Finally, the breathtakingly rapid increase in human diversity triggered a new pattern of variation: regionality. A genetic adaptation to high altitude, for example, was discovered among people living in Tibet; it was a mutation in the gene called EPAS1. This mutation originated 1,000 years ago and spread so fast that it was called the "fastest-evolving gene in the world."* New environments have brought forth cultures and civilizations that are uniquely adaptive; the combination of diverse environments and diverse cultures has led to diverse sets of traits and more rapid evolution. Consequently, human morphology has become more complex and more specifically adaptive.

Foreseeing a diverse future

We often think that evolution happens slowly and gradually—little by little, inconspicuously. Evolution, however, can also take place with terrifying rapidity. We can easily see examples of ultrafast evolution

* Later it was announced that this gene came from the Denisovans.

in agricultural products, farm animals, and pets. They have all been selectively bred to have the form we want, and all the resulting variation in different breeds and crops has occurred in the last 10,000 years. If it's possible for plants and animals, then it's possible for humans too.

In evolution, "advantage" and "benefit" are not inherent and absolute values. A new trait that accidentally happens to be useful for reproduction or adaptation to the environment at that moment is advantageous and beneficial. But that exact trait may be disadvantageous in a different environment. There is no such thing as an absolutely and inherently advantageous trait, nor is there an absolutely and inherently damaging trait.

As biological organisms, we cannot escape the forces and mechanisms of evolution. Humans evolve. But at the same time, humans are unique in that they can influence their own evolutionary trajectory through self-made culture and civilization. None of the traits that humans possess are absolutely advantageous or beneficial, but humans have the ability to apply any trait toward their interest. What's the best thing that humans equipped with such power could do? Perhaps it would be to protect and nurture the environment of the Earth, along with the organisms with which we share it.

What a single human being can do may be minuscule. Together, however, we have explored unknown continents, developed sophisticated cultures, and evolved diverse forms. Taken together, the small actions of each human have added up to monumental achievements for humankind.

EXTRA

EVER-EVOLVING WISDOM TEETH

The third molar (the wisdom tooth) offers a great example of a complex change in variation that is due to advances in medicine. As cooking culture advanced, humans came to prefer soft and well-cooked food. With less chewing, the mandible (jawbone) grew smaller and the gums also receded. Sometimes, upper and lower teeth were not aligned (in other words, there was malocclusion), and the wisdom teeth often had no room to come in. Impacted or crooked wisdom teeth are vulnerable to caries and periodontitis, and when the infection spreads to the whole body, it can even be fatal, or at least create debilitating pain. It is better—that is, selectively advantageous—not to have the wisdom teeth, so it was expected that a mutation for lack of wisdom teeth might spread through human populations. When anthropologists examined prehistoric populations, there was indeed a trend toward an increase in the frequency of third molar agenesis (missing third molars).

With advances in modern dentistry, a new situation is unfolding. Now that wisdom teeth can be extracted when problematic, there is no selective advantage for not making wisdom teeth. In the future, I predict that there will be no increase in the people born with the mutation not to make wisdom teeth. Who knows; there may even be an increase in people *with* wisdom teeth. The point is, we are still evolving, often in unpredictable ways.

EPILOGUE 1

Precious Humanity

In August 2014, a regular reader posted on my Facebook page, inviting me to participate in an online "gratitude relay." It was a fad at the time. A person posts three things to be thankful for, then nominates the next person to continue. Naturally, I tried to think of things to be thankful for that had happened during human evolution. And soon I realized it was not an easy task. There were many things to be thankful for, but all had strings attached. So I started a list.

First on the list was walking upright. When we started walking on two legs, our two arms became free to make tools and to carry things and babies (we can't forget about the babies). New hominin mothers no longer had fur for the babies to cling to. Mothers' two arms held the babies tight and firm over long distances. But bipedalism had a price: lower back pain became a common problem. Those of us with experience know all too well that when your lower back is in pain, there is nothing to be done but lie down, immobilized. There was another price too: an overworked heart. The heart has to send a large volume of blood to the top of the body, against gravity. It is under chronic stress.

The second thing on my list of things to be thankful for was our big brain. Our big brain is literally our namesake: *Homo sapiens* is labeled with and defined by our intelligence. Having a big brain is important for our identity as humans. With big brains and the knowledge contained in them, hominins could process staggering amounts of information, chiefly (at first) to procure a lot of animal fat and protein, which were important in an always competitive and increasingly harsh environment. Big brains allowed hominins to forge and maintain countless beneficial social relationships as well. But to give birth to hominin newborns equipped with brains bigger than our pelvic width, mothers have had to bear bone-breaking pain. Moreover, every childbirth comes with a mortality risk on top of the excruciating pain.

Third, I noted being thankful for our longevity. Hominins that lived longer were able to see the birth of their grandchildren. With the help of grandmothers, hominins could then take care of two or three children at the same time—a previously impossible feat for a species whose children demand such a high parental investment from the moment when they're born. With three overlapping generations, more information could be stored and transmitted to the next generation. The continuation of this longevity trend, however, has led to a situation today in which we have more elderly and fewer youth to care for them, placing a greater economic burden on society.

As I thought about the next two things on my list—agriculture and animal domestication—I felt depressed and perhaps not quite as thankful. Yes, humans became "free" in the sense that they no longer had to rely 100 percent on their foraging skills or the environment to acquire food, and could, in fact, produce as much food as they wanted, even a surplus. Productivity increased, populations exploded, and civilizations expanded. At the same time, however, surplus production led to the privatization of property, and class structures began to develop, along with warfare in which humans killed each other in large numbers for the first time in human history. Moreover, as people abandoned their generational knowledge of ecology and respect for nature, years of crop failures resulted in mass starvation. The concurrent shift to humans and their domesticated animals living in close proximity set up the perfect environment for infectious diseases to jump from animal species to humans, where some became virulent diseases with deadly outcomes. High-density populations became quickly vulnerable to disease epidemics. The price we paid was quite high for the reward of civilization.

As I went down the gratitude list, I could not help but feel the gravity of what these achievements have cost humanity. Perhaps there was nothing in human evolution that we could only be thankful for, with no strings attached. Could it be that gratitude and resentment go hand in hand? Like two sides of a coin? Then I realized something. Maybe the price paid was so high because what we gained was so precious. Nothing in human evolution has come completely for free. Our present situation is a result of the high price we paid throughout our history. And humanity's progress is truly precious.

We must remember, however, that others also paid the price for us. It is not an exaggeration to say that every organism on Earth is paying for us. If we extended our gratitude relay invitation to what Shin-Young Yoon called the "disappearing ones,"* they might respectfully decline. For them, gratitude is an unimaginable luxury, one they cannot afford when they face the planet's most formidable predator: humans.

We have become the strongest and most dangerous form of life in this world. Now it is time for us to take responsibility for the disappearing world that is paying the price for us. Let's take action.

Sang-Hee Lee

* Yoon, Shin-Young. *Asking after the Disappearing Ones* [in Korean]. MID, 2014.

Sang-Hee Lee, working with skull fragments. (© Hee-Chung Lee)

EPILOGUE 2

An Invitation to an Unfamiliar World of Paleoanthropology

This book is based on a collection of essays published in *Gwa Hak Dong A*, a Korean science magazine for the general public, from February 2012 to December 2013, edited and with added materials. I first thought of this series after I worked with Sang-Hee Lee on an article about Neanderthals for a special issue. I had always been fascinated by human evolution and paleoanthropology and was working on a piece about Neanderthals for a special issue of the magazine in March 2011, as it had just been discovered that Neanderthals shared genes with modern humans. I had been pursuing this topic ever since attending a lecture by and communicating with Svante Pääbo from the Max Planck Institute in Germany two years before.

Korea has very few experts in the field of paleoanthropology. I could not find many sources to consult and had just started to look for experts outside Korea. In my search for writings on this topic, I found an article in Korean that reviewed the current state of affairs in human evolution. The author was a paleoanthropologist, a professor at a university in the United States. I sent an email, not entirely sure that I would get a reply—or, if I did, that it would be the reply I wanted—but

I was surprised to receive a prompt and positive response. That was my first encounter with Sang-Hee Lee.

After that we communicated by email, and then talked at length about Neanderthals in a series of international phone calls during the Lunar New Year holiday season in quiet Seoul. The special issue was a success, and I resolved to organize a bigger project with her in the future.

In the following year at an editorial meeting, I came up with the idea of publishing a series of columns on the topic of human evolution. I immediately contacted Lee. I was convinced that this series on paleoanthropology and evolution should not take the typical chronological approach. A sequence starting from where and when the first hominins appeared and moving on to how they evolved through time to finally reach the present seemed too predictable—nothing special. I proposed a unique approach of using an everyday occurrence as a hook to dig deeper into the evolution. I also wanted the writings to reflect Lee's warm and humorous style, so I asked her to write in a conversational style rather than a rigid scholarly style, in order to reach a wider audience. The first several columns were extremely important for securing a lasting readership, so a lot of thinking and care went into them.

When I received the first essay from Lee, I read it over and over again. I worked late into the night, thinking about how best to present the comfortable yet sharp content of the article to the readers. As the editor and organizer of the series, I perhaps put more love into these essays than I would into my own writing. This beautiful collaboration of exchanging inspiration continued for two years, a length of time rarely seen in the world of magazine series in Korea.

As soon as the first column was out, I was approached by *Dong A Il Bo*, a daily newspaper, to start a series for its weekend edition. I edited the essays to work better for the newspaper's readership. They were well received. Simultaneous publication of the series in both the magazine and the newspaper continued for a year, giving us an opportunity to reach a wide range of readers.

These essays, the product of so much love and attention, are finally together in a book. It is now time to reach yet another new readership. I would love to see Lee's inspiring writing reach and inspire many more.

Shin-Young Yoon

Shin-Young Yoon, taking a photo. (© Kyung-Sook Ahn)

APPENDIX 1

Common Questions and Answers about Evolution

"Evolution" has become a word that can be found in many places. It often appears in advertisements. Humans "evolve," refrigerators "evolve," and shampoo "evolves." In ads like these, the word "evolve" means "improve" or "become better." In Korean, the word for "evolution" is derived from a Chinese word consisting of two characters that mean, respectively, "progress" and "become." So the Korean definition, quite literally, is "advancing forward."

Despite the common English usage of "evolved" to mean "better" and "improved," in actuality evolution as the core concept of modern biology has no sense of directionality. It does not imply improvement or things getting better. The definition of "evolution" that biologists have agreed on is simply "changes in the genetic frequency of a population over a long period of time."* Evolution is change, not necessarily progress.

* How long is long? The answer depends on the species, because every species has a different sense of time. For humans, 25 years represents about one generation. But for fruit flies, whose generation spans just 10 days, 25 years covers as many as 910 generations. In other words, in terms of the number of generations, 25 years for fruit flies is equivalent to nearly 23,000 years for humans.

Let's take a closer look at evolution.

Evolutionary theory was surprising and not so surprising

On the one hand, evolutionary theory is quite predictable and sensible. On the other, it is also shockingly new. As with most scientific discoveries, the discovery itself does not have a big impact on our lives. Consider, for example, the debate over geocentrism versus heliocentrism. It may not matter to our everyday life whether we think the sun revolves around the Earth or the Earth revolves around the sun, but obviously there's a big difference.

When heliocentrism was first proposed, the world reacted against the idea, against change. Europeans at the time regarded the world as static, not moving. It was an idealized narrative of our planet as the center of the universe. Perhaps this was perfectly compatible with the medieval idea of a settled life. Since humans are the consummate creation of this world, the world we live in, Earth, should be perfect too. Therefore, Earth keeps its perfect place, and the sun, moon, and stars move about the Earth in search of their perfect place. This was the view of geocentrism. Heliocentrism's proposal that the Earth is the one moving about the sun implied that the perfect Earth where humans, the center of the world, are living, is not perfect after all. It meant the Earth was no longer the center of the universe. It was a dangerous thought.

In its background, evolutionary theory shares some similarity with heliocentrism: chiefly, just as heliocentrism proposed that the Earth is not the center of the universe, evolutionary theory proposes that humans are not the center of the world. Galileo Galilei, who went against the idea that the world we inhabit is at the center of the universe, suffered through torture at the hands of the Inquisition. Likewise, Charles Darwin, who suggested that humans are not the perfect, divinely created pinnacle of all life-forms, suffered through numerous criticisms and attacks on his scholarship. Darwin's theory of natural selection shook the root of the European medieval worldview. It is no wonder that Charles Darwin hesitated for a long time, twenty-three years for that matter, to publish his tome *On the Origin of Species* (1859).

Darwin's idea—that humans adapt like all other creatures—challenged the traditional conviction that humans were at the center of the world, by proposing that they were also part of nature. To embrace this idea meant acknowledging that all things, even animals, change over time; and that no animals, not even humans, are perfect creations of the Christian God. Although the concept of such change was difficult for many to accept, it is compatible with Eastern philosophy. Maybe that's why the opposition to evolution hasn't been as strong in Asia.

Evolution basics

Two fundamental concepts undergird evolution. First, evolution needs raw material, which is genetic variation in the form of mutations. When something different from the existing genes—a novelty—appears, it becomes a variation in the gene pool. Here is a simple example: In a population where everyone has round ears, a mutation occurs that results in pointy ears, like what we imagine an elf has. Before, there was only one version of ear shape, which was round, but now there are two versions for the trait of ear shape: round and pointy.

Second, morphological variation is related to reproductive variation. Continuing with the same example, if the ability to produce offspring is the same for both pointy-eared and round-eared individuals, there will be no evolutionary preference for one ear over the other. But if individuals with pointy ears have a higher probability of leaving offspring than do individuals with round ears, as time passes there will be a higher number of individuals with pointy ears. The frequency of the mutation gene that caused pointy ears will thus increase. It is not a matter of the absolute number of individuals at any given time. Rather, it is the relative prevalence, the "market share," of a particular gene that leads to a particular trait. Evolution is fundamentally a concept of population, and a relative concept.

How can individuals have different probabilities of producing offspring? An individual that is better adapted to the environment and has a higher probability of surviving until adulthood also has a higher probability of leaving behind more offspring than does an individual who dies before reaching adulthood. This distinction is the principle behind "natural selection," the best-known evolutionary mechanism proposed by Darwin.

Sometimes a trait that has no benefit for survival, or even seems to be detrimental for survival, is selected by the opposite sex, and consequently its frequency is increased in subsequent generations. This phenomenon is called sexual selection. A well-known example is a peacock's tail. The long and brilliant tail of a peacock does not help survival, and sometimes it can be a detriment. Its length is cumbersome when agility is called for to escape a predator, and the brilliant color draws attention to and advertises the bird's presence. We would expect such an apparently useless and potentially damaging trait not to help the individual survive and leave offspring, and thus to disappear quickly from the gene pool. But that's not what happened. All peacocks possess this "strange" trait of a large, showy tail.

Darwin proposed the concept of sexual selection to explain this seemingly paradoxical situation. For some reason, peahens select peacocks with the most showy tails to mate with and lay eggs. Perhaps showy tails are a sign of good health. Or perhaps showy tails demonstrate the bravery and resourcefulness of any male peacock that survives predation despite his cumbersome tail. It was the enigmatic choice of the females that selected for the showy tails.

Darwin proposed sexual selection as part of his theory for natural selection. Whether natural or sexual, the basic premise of selection is the same. An individual with an advantageous trait is selected among individuals showing variations of traits that can be selected for or against. The agency of selection may lie with nature or with the opposite sex. Darwin synthesized and developed natural and sexual selection, published as primary mechanisms in evolution theory. Darwin's theory was developed before we had any knowledge of cell biology, inheritance, and genetics; he inferred and deduced his theoretical framework just by observing natural phenomena. How amazing!

The evolution of evolutionary theory

By the 1960s, about 100 years after the publication of Darwin's masterpiece *On the Origin of Species*, selection temporarily lost its explanatory power in the field of evolutionary science. The doubt arose from the seemingly paradoxical situation that most mutations, the source of variation, are not affected by selection. Advantageous mutations spread to all individuals in a population, resulting in no change to the variation, and detrimental mutations are eliminated from the population gene pool, therefore also not changing the variation. The only mutations that remain are neither advantageous nor detrimental, and even those mutations will spread to all individuals in a population or be eliminated, through random processes of time. In other words, mutations that are meaningful for evolution or selection are not observed, while mutations that are observed are neutral in selection.

The neutral theory systematically built upon this idea and argued that only random processes through time or population size, not selection, are the primary mechanism of evolution (see Chapter 18). As a result, big advances were made in population genetics, but at the same time, too few researchers paid attention to the role of selection. In the twenty-first century, selection is again at the center of our collective attention. The theory has come full circle.

The recent advances in epigenetics foretell a new chapter in the evolution of evolutionary theory. The shift would have made Jean-Baptiste Lamarck happy. Lamarck argued for the inheritance of acquired characteristics. He would have explained the long neck of the giraffe as resulting from the animal's efforts to reach leaves in the highest branches of trees. This explanation is different from the Darwinian idea of natural selection, which argues that among many accidental mutations, one in particular caused a long neck, and because this long neck was advantageous in the environment, giraffes that had the long-neck mutation left behind more offspring than did giraffes that lacked that mutation. As we live, our bodies change. Our muscles may become bigger from working out, and our chin may become small and dainty from plastic surgery. But everyone knows that the babies born of people with those bigger muscles or those daintier features will not inherit the big muscles or the small chin. Lamarck's inheritance of acquired characteristics has been stigmatized as the wrong theory until now. But rapid advances made in epigenetics may show how acquired characteristics can be inherited after all.

Common questions about evolution

How is it evolution when it didn't get any better?

A trait that increases in frequency through evolution is definitely selected because it leads to more offspring with that trait than with other traits. Simply being selected, however, does not mean that a trait is necessarily superior or better. Selectively advantageous by natural selection means that the trait happened to be adaptive in a particular environment. With changes in the surrounding conditions, a selectively advantageous trait may turn out to be disadvantageous in the new environment, and many individuals may die out without leaving many offspring. Sexual selection is even more elusive to explain; essentially all we know is that the selected traits are attractive to potential mates. A trait that was once deemed attractive to mates would not necessarily remain attractive forever. Therefore, evolution does not necessarily mean progress.

Where is the missing link?

A common misconception about evolution is the idea of a "missing link." The missing link was a popular concept that sprang up in the early years of evolutionary theory. The core idea was that if evolution really happened, there should be missing links to fill in the gaps between fossils when they were arranged in order of a particular morphological change. This position argues that fossil data are spotty and the trajectory of change is smooth and linear. Therefore, since fossils haven't been found to fill in the missing links of the evolutionary trajectory, evolutionary theory is suspect. The obvious first objection to this concept is that fossilized remains are difficult to come by, as they must be preserved under a certain set of rare conditions. Moreover, a variety of theories that support evolution, including the development of the punctuated equilibrium model by Stephen Jay Gould and Niles Eldredge, argue that changes occur not gradually and smoothly, but rather in concentrated spurts followed by long periods of stasis. The missing-link idea has so far proved unconvincing.

If we evolved from monkeys, there should be monkeys that are evolving into humans even now. Where are they?

Wait, let's make sure we're on the same page first. Humans evolved from apes, not monkeys; it might sound like a small distinction, but it's important to be clear. Many people think of chimpanzees as monkeys. Chimpanzees are apes, not monkeys. The easiest distinction between apes and monkeys is whether or not they have tails. Monkeys have tails, apes do not; simple as that. Ironically, the last ape to have its genome sequenced was a gibbon, whose Korean name translates to "long-armed monkey." With some of us still calling an ape a monkey, the distinction continues to be difficult to grasp. Quite unfortunate.

Whether people are talking about monkeys or apes, though, this question comes from a deeper misconception: the idea that all organisms in this world are evolving to attain the highest place in the biological chain of being, currently occupied by humans. According to this idea, all organisms can be placed in a sequence ordered by the distance from the top, arranged in terms of similarity, with animals higher on the chain closer to humans and lower animals farther from humans. Likewise, "lower" animals are always striving to be "higher" animals, with humans as the proposed end goal.

By this reasoning, the idea persists that there are apes, or even monkeys, that are right now striving to become humans. Monkeys aren't exactly the chopped liver of evolutionary development, though; they've had their own respectable evolutionary history up to now, wouldn't you say? Maybe they don't want to become humans! Joking aside, the practice of arranging all organisms in one linear sequence, placing humans at one end and the rest in the order of how different or similar they appear in comparison with humans is outdated and no longer recognized by the field of modern biology. Even the lowliest organism—a tapeworm, let's say—is an evolutionary triumph, simply by its existence today.

APPENDIX 2

Overview of Hominin Evolution

When did ancestral humans start and where? In order to answer that question, we need to ask where and when humans diverged from the common ancestors we share with our closest living relative species, the chimpanzee. According to research in molecular biology, modern human ancestors and chimpanzee ancestors diverged between 8 and 5 million years ago, sometime during the Miocene, in Africa. The exact point in time and the biological circumstances of our divergence, however, are not well known. This uncertainty is due to the dearth in fossil data from that critical time period. Several fossils found within the last ten years have been claimed to be the earliest ancestor for the human lineage. Among *Orrorin tugenensis, Sahelanthropus tchadensis, Ardipithecus kadabba,* and *Ardipithecus ramidus,* it is not clear whether they are examples of the first hominins, or if they are prehominins, belonging to the lineage before the divergence of humans and chimpanzees.

The fossil species that are certain to come after the point of divergence between the human and the chimpanzee lineages (therefore, certainly hominins) are australopithecines from 3–4 million years ago, during the Pliocene. Well known among them are *Australopithe-*

cus anamensis, Australopithecus afarensis, Australopithecus/Paranthropus boisei, and *Australopithecus/Paranthropus aethiopicus* from eastern Africa; and *Australopithecus africanus* and *Australopithecus/Paranthropus robustus* from southern Africa. In addition, discoveries of more australopithecine species, such as *Australopithecus garhi, Australopithecus bahrelghazali, Australopithecus sediba*, and *Kenyanthropus platyops* were also announced in the 1990s.* Considering that these species are based on fossils from a single site, we need to wait and see whether they will be validated in time as separate species.

Early hominins represented by *Australopithecus anamensis* and *Australopithecus afarensis* look quite similar, in cranial capacity (brain size) and in cranial and dental morphology, to apes such as chimpanzees and gorillas. Except they walked upright as bipeds, like we do. The bipedalism of the earliest hominins probably included a climbing adaptation for trees and other terrain, unlike the obligate bipedalism of later hominins.

After *Australopithecus afarensis*, ancestral hominins that were spread throughout southern and eastern Africa show various adaptive strategies to survive in the colder and drier environment. Australopithecines of the Late Pliocene specialized in eating and digesting huge amounts of plant-based food of poor nutritional quality. This adaptive strategy is reflected in the overly developed masticatory (chewing) characteristics in their anatomy. For example, *Australopithecus/Paranthropus aethiopicus* had molars as big as those of present-day gorillas,

* Of these, *Australopithecus sediba* is the one gaining the most support, in the form of substantial amounts of fossil data from continued excavations. In addition, the research team studying *Australopithecus sediba* broke with the longtime tradition of sharing data only within an "in-group" and opened access to their new data, inviting young scholars in the beginning stages of their careers to do original research on the fossils. The leader of the team, Lee Berger, did even more to revolutionize the field, by granting transparency and free access to data related to the discovery of and research on *Homo naledi*.

but their body was only one-fourth the size. This means they had to eat as much food as gorillas do, but barely maintained a body size one-fourth that of gorillas; in other words, the food they ate was of such poor quality that they had to eat a lot to get the same amount of nutrition and calories.

In contrast, the genus *Homo*, our lineage, which appeared during the Late Pliocene and Early Pleistocene, relied a little more on animal-based food (meat) and had a bigger brain. Broadly speaking, there were two ways to get animal-based food: one was to consume carcasses left behind by other predatory animals, and the other was to directly hunt live game. The lineages that collected food from carcasses, *Homo habilis* and *Homo rudolfensis*, used stone tools to crack open large limb bones and to extract the precious bone marrow.

Another species of the genus *Homo*, *Homo erectus* (sometimes called *Homo ergaster*), is more closely related to modern humans than is either *Homo habilis* or *Homo rudolfensis*. *Homo erectus* made and used stone tools as an adaptation for hunting live game and, with the high-protein, high-fat diet made possible by hunting, could afford to have a bigger brain and bigger body. *Homo erectus* was active during the day, avoiding the competition for live game with other, mostly nocturnal, predators. This change was possible because of the new physiological adaptation of thermoregulation through sweating and the development of longer limbs without fur. Our furless bodies, in turn, led to the advantageous adaptation of melanin synthesis to ward off the dangers of ultraviolet radiation during the scorching heat of daytime in equatorial Africa. Ancestral humans had a large body and a large brain, and most likely became naked (furless) and then dark-skinned as a result. They walked upright, in a style almost the same as that of modern humans.

The genus *Homo* was the first hominin lineage that spread out of Africa, as indicated by many fossil discoveries in Europe and Asia. Why and how did hominins leave Africa and spread worldwide? The

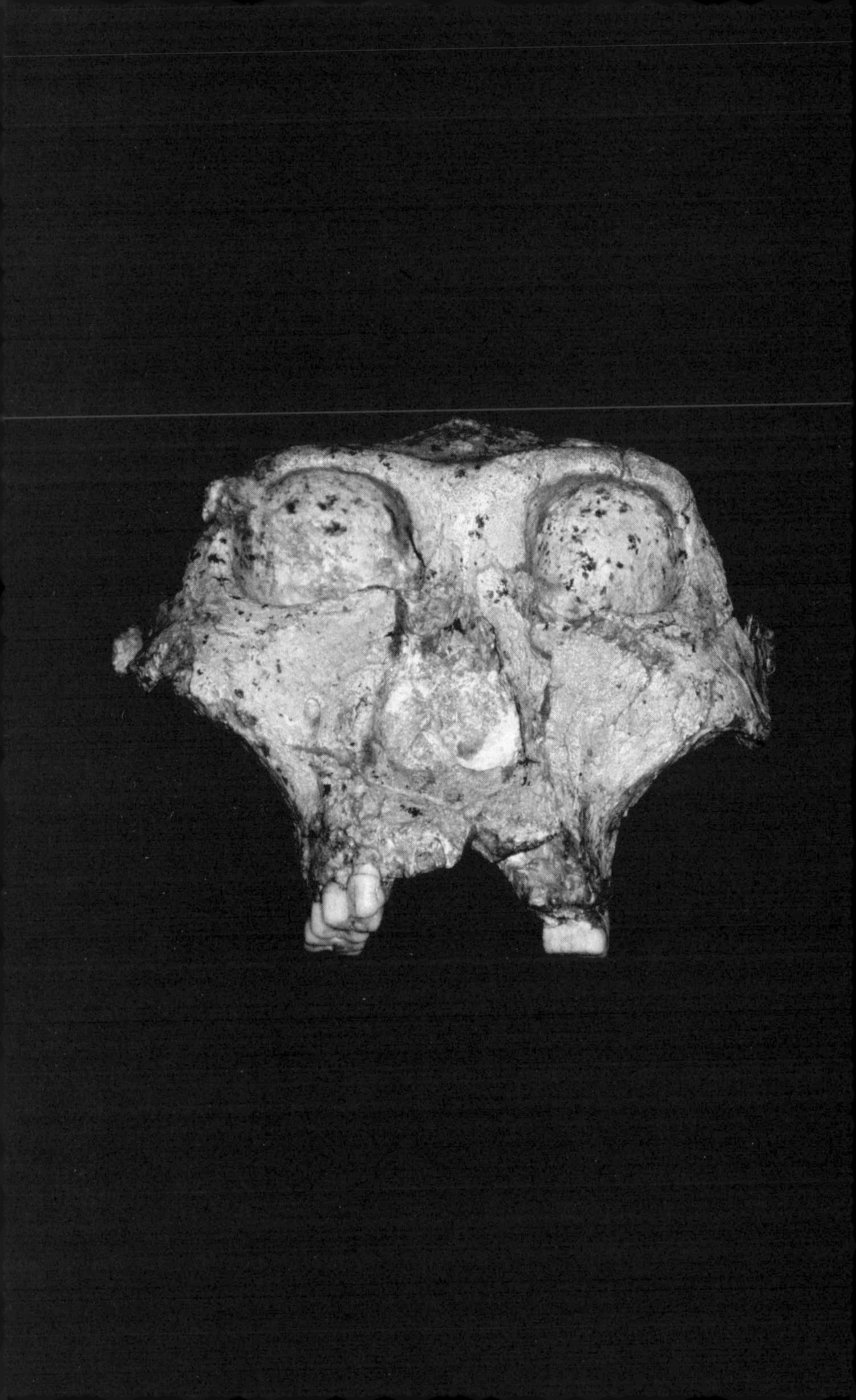

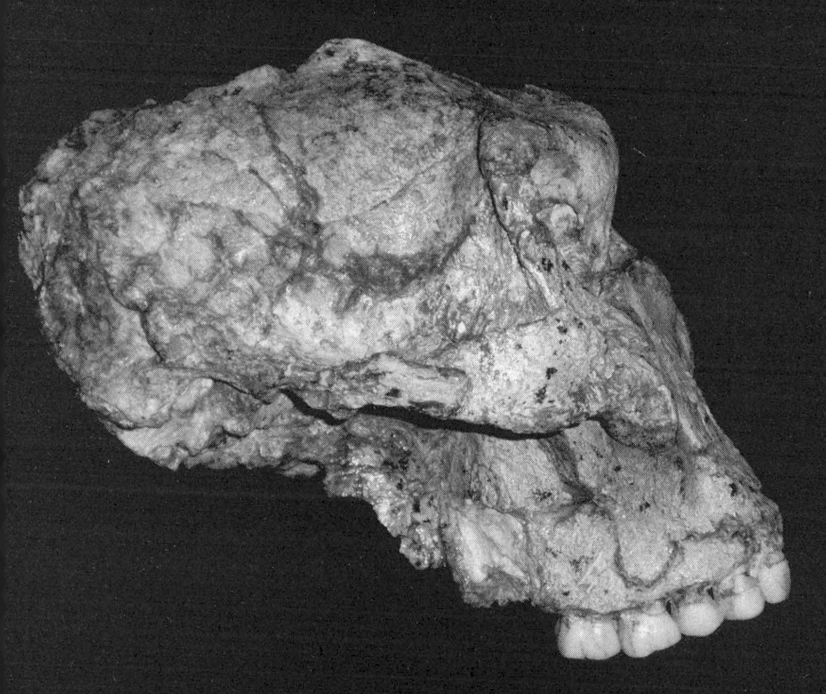

Two views of fossil skull of *Australopithecus/Paranthropus robustus*, discovered in Swartkrans, South Africa. (© Milford Wolpoff)

mainstream model says that the first species of the genus *Homo* originated in Africa about 1.8–2.3 million years ago. Being well adapted to hunting by virtue of their big brain and big body, they followed big game into Europe and Asia about 700,000–800,000 years ago, when big game left Africa because of climate change.

Recent discoveries, however, demand that we reexamine this model. The oldest *Homo* in Europe (*Homo georgicus,* discovered in Dmanisi, Georgia) and the oldest *Homo* in Asia (*Homo erectus,* found in Java, Indonesia, also known as Java Man) are now known to be as old as 1.8 million years, which is almost as old as the earliest *Homo erectus* in Africa. Accordingly, some scholars argue for an Asian, instead of African, origin of the genus *Homo. Homo georgicus* from Georgia had a small brain and small body, in direct challenge to the mainstream, hunting hypothesis that was predominant until recently. So far, however, no new hypothesis has been convincing enough to take the place of the mainstream model.

How did hominins branch out from Africa into Eurasia? The worldwide spread of hominins was not a purposeful movement by a nation. Instead, it was most likely an organic process resulting from population increase and population pressure or perhaps even following prey migratory patterns. If so, we can posit that an increase in fertility or a decrease in mortality preceded the spread. We often worry about population explosion during modernization, but in early hominins it's worth keeping in mind that it was not easy to have an increase in fertility among mobile populations, simply because it is not easy to move around with more than one child, considering everything else that has to be moved as well.

Among modern humans, a child is considered more or less independent at six to seven years of age, so it would have been best in these early hominins to wait about five or six years between births so that an older child could move around independently by the time the second child was born. Indeed, we see that for a mobile population such as the !Kung, the interbirth interval (the interval between two births) is about five years. A spread into new regions that was prompted by an increase in fertility would therefore suggest that the interbirth interval had become shorter. And it would further indicate a social support system in which the responsibility of taking care of more than two barely walking children could be shared. This social support may have come from the father (according to the provisioning hypothesis, or Lovejoy model) or the maternal grandmother (according to the grandmother hypothesis), but the debates are ongoing and far from settled.

As hominins spread to Eurasia, regional populations with regional traits started to appear in the Middle Pleistocene. Species names are often given to these populations, but opinions vary as to the validity of these species. Currently, those that are recognized as species in Europe are *Homo heidelbergensis* and *Homo neanderthalensis*. In Africa, the place of origin for the genus *Homo*, there are *Homo erectus/ergaster* and *Homo heidelbergensis* (hypothesized to have come back to Africa from Europe). In Asia, there is *Homo erectus*. Additional species from the Middle Pleistocene are based on a couple of sites or a couple of fossils, such as *Homo cepranensis* (based on the Ceprano fossil from Italy), *Homo antecessor* (from the Atapuerca site in Spain), *Homo floresiensis* (from Flores, Indonesia), and *Homo rhodesiensis* (from Kabwe, Zambia). Two other examples of note are *Homo georgicus* from the Early Pleistocene and, recently, the Denisovans, from Denisova, Russia.

Are these all equally valid species? In paleoanthropology, there has been a tendency to announce a new species whenever a fossil is found in a new region. But it is unclear whether these fossils with new species names are really new biological species. Especially if the species is found in only one site, it is likely to be subsumed sooner or later under another species with a larger regional distribution. A famous example is Peking Man, which was discovered in the Zhoukoudian cave, China; announced as *Sinanthropus pekinensis*; later reclassified as *Pithecanthropus erectus* along with Java Man from Java, Indonesia; and then reclassified again, as *Homo erectus,* when *Pithecanthropus* was reclassified as *Homo.* This is only one of many such examples.

How are all the species or populations from the genus *Homo* related to modern humans, *Homo sapiens*? There are two possible answers, depending on how the origin of *Homo sapiens* is understood. One is the complete replacement model of modern humans. According to this model, *Homo sapiens* originated as a new species "recently" (from the perspective of human evolutionary history), about 200,000 years ago, in Africa. Because *Homo sapiens* spread from Africa into Eurasia only recently as a new species different from other species, it had no admixture with the indigenous hominins living in those regions. Equipped with superior culture and language, *Homo sapiens* outcompeted the other indigenous populations, which all went extinct. Herto, a hominin fossil found in Ethiopia, is the type specimen for the new subspecies of *Homo sapiens* called *Homo sapiens idaltu.* This population is the one that spread out of Africa into the world, with no connection to any archaic hominins.

The other model, called the multiregional evolution model, does not regard modern humans as a new species that originated from one place. Modern humans do not have a single ancestry. Populations from all regions, and from all time periods, have continued to exchange culture and genes for the last 2 million years. The populations that went extinct or arose anew in the meantime are all part of the same species; therefore, there has been no new species in human evolution since the Pleistocene.

The biological definition of species lies in the ability to interbreed and produce viable offspring. If populations continue to exchange genes, they are of the same species. If such interbreeding has gone on for the last 2 million years, then *Homo sapiens* originated at least 2 million years ago. The morphological traits that we observe in modern humans did not originate in one single place, but in many different places. Traits that are advantageous on a global scale will spread worldwide; those that are advantageous for local adaptation will be just regional characteristics. Examples of globally advantageous traits include gracilization (a trend toward a softer, less robust form) and encephalization (an increase in cranial capacity), morphological traits that appear in all regions of the world. Regionally advantageous traits are exemplified by the shovel-shaped incisors that are common in modern Asian populations and can be observed in the early-hominin fossils found in China. Any one of the traits commonly observed in both Neanderthals and modern Europeans are also regional characteristics, such as the midfacial prognathism (jutting out) that results from their enlarged maxillary sinuses, which force the middle portion of the face to protrude forward.

The multiregional evolution model has traditionally found strong support in fossil data. In contrast, the complete replacement model started to gain support in the 1990s, when geneticists showed through research on modern-human genes that modern humans do not have a long history, and that our place of origin is Africa. Since then, as molecular biology gained in stature, so did the complete replacement model. In particular, research published between 1997 and 2000 showing that ancient DNA extracted from Neanderthal fossils was quite different from that of modern humans—and therefore implying that Neanderthals made no contribution to the genetic makeup of modern humans—gained strong (almost universal) support.

In the last several years, however, there has been a reversal: research in population genetics and the sequencing of the full Neanderthal genome in 2010 showed that Neanderthals did contribute to the genetics of modern humans. Now the majority of the field is distancing itself from the complete replacement model.

The problem with the multiregional evolution model lies in the logical conclusion that *Homo sapiens* is as old as 2 million years. If all populations continued to exchange genes through time and space, they all belong to the same species by definition of a biological species because they share a specific gene pool. If all the hominins that appeared after *Homo erectus* originated in Africa belonged to the same species, then *Homo erectus* and *Homo sapiens* would ultimately belong to the same species. And following the established rule of naming species, *Homo erectus* would be reclassified as *Homo sapiens*. In that case, *Homo erectus*, a species name that has been used for more than 100 years, would become a population name rather than a species name. And this would happen with all species of the genus *Homo*, except *Homo habilis*, which is understood as a separate species, with some paleoanthropologists arguing that it belongs to the *Australopithecus* genus rather than the *Homo* genus. All this makes sense logically, but it is quite difficult to swallow from the conventional point of view. If we followed every one of these suggestions, then the genus *Homo* would be a genus with a single species, *Homo sapiens*!

Paleoanthropological research began a new chapter in the twenty-first century. Now we even have an ancestral human population—the Denisovans—that exists only as DNA, without a substantial fossil representation. As the technology for extracting ancient DNA without contaminating it advances and the cost of such technology decreases, genetics will continue to have greater and greater impact in paleoanthropology, becoming as important for data as fossils are, if not more so. But new fossils continue to be discovered. As the technology to collect and analyze these new fossils improves, so will our data and research efforts. Through it all, we will continue to ask and seek answers to fundamental questions. Where did humans come from? What was the path humans took to get where we are today? And where will our human path lead?

Further Reading

Introduction: Let's Take a Journey Together

Books

Bryson, Bill. *The Lost Continent: Travels in Small-Town America.* Secker, 1989.

Steinbeck, John. *Travels with Charley: In Search of America.* Viking, 1962.

1. Are We Cannibals?

Books

Arens, William. *The Man-Eating Myth: Anthropology and Anthropophagy.* Oxford University Press, 1979.

White, Tim D. *Prehistoric Cannibalism: At Mancos 5MTUMR-2346.* Princeton University Press, 1992.

Articles

Defleur, Alban, Tim White, Patricia Valensi, Ludovic Slimak, and Évelyne Crégut-Bonnoure. "Neanderthal Cannibalism at Moula-Guercy, Ardèche, France." *Science* 286, no. 5437 (1999): 128–31.

Gajdusek, D. Carleton. "Unconventional Viruses and the Origin and Disappearance of Kuru." *Science* 197, no. 4307 (1977): 943–60.

Marlar, Richard A., Banks L. Leonard, Brian R. Billman, Patricia M. Lambert, and Jennifer E. Marlar. "Biochemical Evidence of Cannibalism at a Prehistoric Puebloan Site in Southwestern Colorado." *Nature* 407, no. 6800 (2000): 74–78.

Rougier, Hélène, Isabelle Crevecoeur, Cédric Beauval, Cosimo Posth, Damien Flas, Christoph Wissing, Anja Furtwängler, et al. "Neandertal Cannibalism and Neandertal Bones Used as Tools in Northern Europe." *Scientific Reports* 6 (2016): 29005.

Russell, Mary D. "Mortuary Practices at the Krapina Neandertal Site." *American Journal of Physical Anthropology* 72, no. 3 (1987): 381–97.

White, Tim D. "Once Were Cannibals." *Scientific American* 265, no. 2 (2001): 58–65.

2. The Birth of Fatherhood

Books

Gray, Peter B., and Kermyt G. Anderson. *Fatherhood: Evolution and Human Paternal Behavior.* Harvard University Press, 2012.

Hager, Lori D., ed. *Women in Human Evolution.* Routledge, 1997.

Hrdy, Sarah Blaffer. *The Woman That Never Evolved.* Harvard University Press, 1999.

Lee, R. B., and I. DeVore, eds. *Man the Hunter.* Aldine, 1968.

Articles

Bribiescas, Richard G. "Reproductive Ecology and Life History of the Human Male." *Yearbook of Physical Anthropology* 44 (2001): 148–76.

Gray, Peter B. "Evolution and Human Sexuality." *American Journal of Physical Anthropology* 152, no. S57 (2013): 94–118.

Lovejoy, C. Owen. "The Origin of Man." *Science* 211, no. 4480 (1981): 341–50.

3. Who Were the First Hominin Ancestors?

Books

Tattersall, Ian. *Masters of the Planet: The Search for Our Human Origins.* St. Martin's Griffin, 2013.

Articles

Asfaw, Berhane, Tim D. White, C. Owen Lovejoy, Bruce Latimer, Scott Simpson, and Gen Suwa. "*Australopithecus garhi*: A New Species of Early Hominid from Ethiopia." *Science* 284, no. 5414 (1999): 629–35.

Brunet, Michel, Franck Guy, David R. Pilbeam, Hassane Taïsso Mackaye, Andossa Likius, Djimdoumalbaye Ahounta, Alain Beauvilain, et al. "A New Hominid from the Upper Miocene of Chad, Central Africa." *Nature* 418, no. 6894 (2002): 145–51.

Dart, Raymond A. "*Australopithecus africanus*: The Man-Ape of South Africa." *Nature* 115, no. 2884 (1925): 195–99.

Gibbons, Ann. "In Search of the First Hominids." *Science* 295 , no. 5558(2002): 1214–19.

Johanson, Donald C., and Tim D. White. "A Systematic Assessment of Early African Hominids." *Science* 203, no. 4378 (1979): 321–30.

Leakey, Meave G., Craig S. Feibel, Ian McDougall, Carol Ward, and Alan Walker. "New Specimens and Confirmation of an Early Age for *Australopithecus anamensis*." *Nature* 393, no. 6680 (1998): 62–66.

Leakey, Meave G., and Alan C. Walker. "Early Hominid Fossils from Africa." *Scientific American* 276, no. 6 (1997): 74–79.

Sarich, Vincent M., and Allan C. Wilson. "Immunological Time Scale for Hominid Evolution." *Science* 158, no. 3805 (1967): 1200–3.

Senut, Brigitte, Martin H. L. Pickford, Dominique Gommery, P. Mein, K. Cheboi, and Yves Coppens. "First Hominid from the Miocene (Lukeino Formation, Kenya)." *Comptes Rendus de l'Académie des Sciences Paris* 332, no. 2 (2001): 137–44.

White, Tim D., Berhane Asfaw, Yonas Beyene, Yohannes Haile-Selassie, C. Owen Lovejoy, Gen Suwa, and Giday WoldeGabriel. "*Ardipithecus ramidus* and the Paleobiology of Early Hominids." *Science* 326, no. 5949 (2009): 64, 75–86.

Wong, Kate. "An Ancestor to Call Our Own." *Scientific American* 288, no. 1 (2003): 54–63.

4. Big-Brained Babies Give Moms Big Grief

Books

Trevathan, Wenda R. *Human Birth: An Evolutionary Perspective.* Aldine, 1987.

Articles

Gibbons, Ann. "The Birth of Childhood." *Science* 322, no. 5904 (2008): 1040–43.

Ponce de León, Marcia S., Lubov Golovanova, Vladimir Doronichev, Galina Romanova, Takeru Akazawa, Osamu Kondo, Hajime Ishida, and Christoph P. E. Zollikofer. "Neanderthal Brain Size at Birth Provides Insights into the Evolution of Human Life History." *Proceedings of the National Academy of Sciences of the USA* 105, no. 37 (2008): 13764–68.

Rosenberg, Karen R., and Wenda R. Trevathan. "Bipedalism and Human Birth: The Obstetrical Dilemma Revisited." *Evolutionary Anthropology* 4, no. 5 (1996): 161–68.

Rosenberg, Karen R., and Wenda R. Trevathan. "The Evolution of Human Birth." *Scientific American* 285, no. 5 (2001): 76–81.

Simpson, Scott W., Jay Quade, Naomi E. Levin, Robert Butler, Guillaume Dupont-Nivet, Melanie Everett, and Sileshi Semaw. "A Female *Homo erectus* Pelvis from Gona, Ethiopia." *Science* 322, no. 5904 (2008): 1089–92.

5. Meat Lovers R Us

Books

Lee, R. B., and I. DeVore, eds. *Man the Hunter.* Aldine, 1968.

Stanford, Craig B. *The Hunting Apes: Meat Eating and the Origins of Human Behavior.* Princeton University Press, 1999.

Articles

Finch, Caleb E., and Craig B. Stanford. "Meat-Adaptive Genes and the Evolution of Slower Aging in Humans." *Quarterly Review of Biology* 79, no. 1 (2004): 2–50.

Speth, John D. "Thoughts about Hunting: Some Things We Know and Some Things We Don't Know." *Quaternary International* 297 (2013): 176–85.

Walker, Alan, M. R. Zimmerman, and R. E. F. Leakey. "A Possible Case of Hypervitaminosis A in *Homo erectus.*" *Nature* 296, no. 5854 (1982): 248–50.

6. Got Milk?

Books

Wiley, Andrea S. *Re-imagining Milk: Cultural and Biological Perspectives.* Routledge, 2010.

Articles

Beja-Pereira, Albano, Gordon Luikart, Phillip R. England, Daniel G. Bradley, Oliver C. Jann, Giorgio Bertorelle, Andrew T. Chamberlain, et al. "Gene-Culture Coevolution between Cattle Milk Protein Genes and Human Lactase Genes." *Nature Genetics* 35, no. 4 (2003): 311–13.

Burger, J., M. Kirchner, B. Bramanti, W. Haak, and M. G. Thomas. "Absence of the Lactase-Persistence-Associated Allele in Early

Neolithic Europeans." *Proceedings of the National Academy of Sciences of the USA* 104, no. 10 (2007): 3736–41.

Enattah, Nabil Sabri, Tine G. K. Jensen, Mette Nielsen, Rikke Lewinski, Mikko Kuokkanen, Heli Rasinpera, Hatem El-Shanti, et al. "Independent Introduction of Two Lactase-Persistence Alleles into Human Populations Reflects Different History of Adaptation to Milk Culture." *American Journal of Human Genetics* 82, no. 1 (2008): 57–72.

Tishkoff, Sarah A., Floyd A. Reed, Alessia Ranciaro, Benjamin F. Voight, Courtney C. Babbitt, Jesse S. Silverman, Kweli Powell, et al. "Convergent Adaptation of Human Lactase Persistence in Africa and Europe." *Nature Genetics* 39, no. 1 (2007): 31–40.

Wiley, Andrea S. " 'Drink Milk for Fitness': The Cultural Politics of Human Biological Variation and Milk Consumption in the United States." *American Anthropologist* 106, no. 3 (2004): 506–17.

7. A Gene for Snow White

Books

Jablonski, Nina G. *Skin: A Natural History*. University of California Press, 2006.

Articles

Jablonski, Nina G., and George Chaplin. "Skin Deep." *Scientific American* 287, no. 4 (2002): 74–81.

Mathieson, Iain, Iosif Lazaridis, Nadin Rohland, Swapan Mallick, Nick Patterson, Songül Alpaslan Roodenberg, Eadaoin Harney, et al. "Genome-wide Patterns of Selection in 230 Ancient Eurasians." *Nature* 528, no. 7583 (2015): 499–503.

Myles, Sean, Mehmet Somel, Kun Tang, Janet Kelso, and Mark Stoneking. "Identifying Genes Underlying Skin Pigmentation Differences among Human Populations." *Human Genetics* 120, no. 5 (2006): 613–21.

Rana, Brinda K., David Hewett-Emmett, Li Jin, Benny H.-J. Chang, Naymkhishing Sambuughin, Marie Lin, Scott Watkins, et al. "High Polymorphism at the Human Melanocortin I Receptor Locus." *Genetics* 151, no. 4 (1999): 1547–57.

Wilde, Sandra, Adrian Timpson, Karola Kirsanow, Elke Kaiser, Manfred Kayser, Martina Unterländer, Nina Hollfelder, et al. "Direct Evidence for Positive Selection of Skin, Hair, and Eye Pigmentation in Europeans during the Last 5,000 y." *Proceedings of the National Academy of Sciences of the USA* 111, no. 13 (2014): 4832–37.

8. Granny Is an Artist

Books

Hawkes, Kristen, and Richard R. Paine, eds. *The Evolution of Human Life History*. School of American Research Press, 2006.

Articles

Caspari, Rachel. "The Evolution of Grandparents." *Scientific American*, 305, no. 2 (2011): 44–49.

Caspari, Rachel E., and Sang-Hee Lee. "Is Human Longevity a Consequence of Cultural Change or Modern Biology?" *American Journal of Physical Anthropology* 129, no. 4 (2006): 512–17.

Caspari, Rachel E., and Sang-Hee Lee. "Older Age Becomes Common Late in Human Evolution." *Proceedings of the National Academy of Sciences of the USA* 101, no. 30 (2004): 10895–900.

Hawkes, Kristen. "Grandmothers and the Evolution of Human Longevity." *American Journal of Human Biology* 15, no. 3 (2003): 380–400.

Hawkes, Kristen, James F. O'Connell, Nicholas G. Blurton Jones, Helen Perich Alvarez, and Eric L. Charnov. "Grandmothering,

Menopause, and the Evolution of Human Life Histories." *Proceedings of the National Academy of Sciences of the USA* 95, no. 3 (1998): 1336–39.

Kaplan, Hillard S., and Arthur J. Robson. "The Emergence of Humans: The Coevolution of Intelligence and Longevity with Intergenerational Transfers." *Proceedings of the National Academy of Sciences of the USA* 99, no. 15 (2002): 10221–26.

Lee, Ronald D. "Rethinking the Evolutionary Theory of Aging: Transfers, Not Births, Shape Senescence in Social Species." *Proceedings of the National Academy of Sciences of the USA* 100, no. 16 (2003): 9637–42.

Lee, Sang-Hee. "Human Longevity and World Population." In *21st Century Anthropology: A Reference Handbook*, edited by H. James Birx, 970–76. Sage, 2010.

9. Did Farming Bring Prosperity?

Books

Cohen, Mark Nathan, and George J. Armelagos, eds. *Paleopathology at the Origins of Agriculture*. Academic Press, 1984.

Diamond, Jared. *Guns, Germs, and Steel*. W. W. Norton, 1997.

Articles

Armelagos, George J. "Health and Disease in Prehistoric Populations in Transition." In *Disease in Populations in Transition: Anthropological and Epidemiological Perspectives*, edited by A. C. Swedlund and George J. Armelagos, 127–44. Begin and Garvey, 1990.

Armelagos, George J., Alan H. Goodman, and Kenneth H. Jacobs. "The Origins of Agriculture: Population Growth during a Period of Declining Health." *Population & Environment* 13, no. 1 (1991): 9–22.

Bellwood, Peter S. "Early Agriculturalist Diasporas? Farming,

Languages, and Genes." *Annual Review of Anthropology* 30 (2001): 181–207.

Bocquet-Appel, Jean-Pierre, and Stephan Naji. "Testing the Hypothesis of a Worldwide Neolithic Demographic Transition: Corroboration from American Cemeteries." *Current Anthropology* 47, no. 2 (2006): 341–65.

Larsen, Clark Spencer. "Biological Changes in Human Populations with Agriculture." *Annual Review of Anthropology* 24 (1995): 185–213.

Marlowe, Frank. "Hunter-Gatherers and Human Evolution." *Evolutionary Anthropology* 14, no. 2 (2005): 54–67.

10. Peking Man and the Yakuza

Books

Boaz, Noel T., and Russell L. Ciochon. *Dragon Bone Hill: An Ice-Age Saga of Homo erectus*. Oxford University Press, 2004.

Rightmire, G. Philip. *The Evolution of Homo erectus: Comparative Anatomical Studies of an Extinct Human Species*. Cambridge University Press, 1990.

Articles

Antón, Susan C. "Natural History of *Homo erectus*." *American Journal of Physical Anthropology, Supplement: Yearbook of Physical Anthropology* 122, no. S37 (2003): 126–70.

Berger, Lee R., Wu Liu, and Xiujie Wu. "Investigation of a Credible Report by a US Marine on the Location of the Missing Peking Man Fossils." *South African Journal of Science*, no. 108 (2012): 3–5.

Shen, Guanjun, Xing Gao, Bin Gao, and Darryl E. Granger. "Age of Zhoukoudian *Homo erectus* Determined with 26Al/10Be Burial Dating." *Nature* 458, no. 7235 (2009): 198–200.

Weidenreich, Franz. *The Dentition of Sinanthropus pekinensis: A Comparative Odontography of the Hominids*. Palaeontologia Sinica, New Series D, no. 1. 1937.

Weidenreich, Franz. *The Skull of Sinanthropus pekinensis: A Comparative Study of a Primitive Hominid Skull*. Palaeontologia Sinica, New Series D, no. 10. 1943.

Wu, Xiujie, Lynne A. Schepartz, and Christopher J. Norton. "Morphological and Morphometric Analysis of Variation in the Zhoukoudian *Homo erectus* Brain Endocasts." *Quaternary International* 211, no. 1–2 (2010): 4–13.

11. Asia Challenges Africa's Stronghold on the Birthplace of Humanity

Books

Shipman, Pat. *The Man Who Found the Missing Link: Eugene Dubois and His Lifelong Quest to Prove Darwin Right*. Harvard University Press, 2001.

Spencer, Frank. *Piltdown: A Scientific Forgery*. Oxford University Press, 1990.

Swisher, Carl C., III, Garniss H. Curtis, and Roger Lewin. *Java Man: How Two Geologists' Dramatic Discoveries Changed Our Understanding of the Evolutionary Path to Modern Humans*. Scribner, 2000.

Articles

Dart, Raymond A. "*Australopithecus africanus*: The Man-Ape of South Africa." *Nature* 115, no. 2884 (1925): 195–99.

Dennell, Robin W. "Human Migration and Occupation of Eurasia." *Episodes* 31, no. 2 (2008): 207–10.

Gabunia, Leo, Abesalom Vekua, David Lordkipanidze, Carl C. Swisher III, Reid Ferring, Antje Justus, Medea Nioradze, et al.

"Earliest Pleistocene Hominid Cranial Remains from Dmanisi, Republic of Georgia: Taxonomy, Geological Setting, and Age." *Science* 288, no. 5468 (2000): 1019–25.

Kaifu, Yousuke, and Masaki Fujita. "Fossil Record of Early Modern Humans in East Asia." *Quaternary International* 248 (2012): 2–11.

Lordkipanidze, David, Marcia S. Ponce de León, Ann Margvelashvili, Yoel Rak, G. Philip Rightmire, Abesalom Vekua, and Christoph P. E. Zollikofer. "A Complete Skull from Dmanisi, Georgia, and the Evolutionary Biology of Early *Homo*." *Science* 342, no. 6156 (2013): 326–31.

Wong, Kate. "Stranger in a New Land." *Scientific American* 289, no. 5 (2003): 74–83.

12. Cooperation Connects You and Me

Books

Axelrod, Robert. *The Evolution of Cooperation*. Basic Books, 1984.

Solecki, Ralph S. *Shanidar: The First Flower People*. Knopf, 1971.

Wilson, Edward O. *On Human Nature*. Harvard University Press, 1978.

Wilson, Edward O. *Sociobiology: The New Synthesis*. Belknap Press, 1975.

Articles

Hamilton, W. D. "The Evolution of Altruistic Behavior." *American Naturalist* 97, no. 896 (1963): 354–56.

Lee, Ronald D. "Rethinking the Evolutionary Theory of Aging: Transfers, Not Births, Shape Senescence in Social Species." *Proceedings of the National Academy of Sciences of the USA* 100, no. 16 (2003): 9637–42.

Lordkipanidze, David, Abesalom Vekua, Reid Ferring, G. Philip Rightmire, Jordi Agusti, Gocha Kiladze, Aleksander Mouskhelishvili, et al. "The Earliest Toothless Hominin Skull." *Nature* 434 (2005): 717–718.

Nowak, Martin A., and Karl Sigmund. "Evolution of Indirect Reciprocity." *Nature* 437, no. 7063 (2005): 1291–98.

13. King Kong

Books

Ciochon, Russell L., John W. Olsen, and Jamie James. *Other Origins: The Search for the Giant Ape in Human Prehistory.* Bantam, 1990.

Weidenreich, Franz. *Apes, Giants, and Man.* University of Chicago Press, 1946.

Articles

Lee, Sang-Hee, Jessica W. Cade, and Yinyun Zhang. Patterns of Sexual Dimorphism in *Gigantopithecus blacki* Dentition. *American Journal of Physical Anthropology* 144, no. S52 (2011): 197.

Pei, Wen-Chung. "Giant Ape's Jaw Bone Discovered in China." *American Anthropologist* 59, no. 5 (1957): 834–38.

Simons, Elwyn L., and Peter C. Ettel. "Gigantopithecus." *Scientific American* 222, no. 1 (1970): 76–85.

Von Koenigswald, G. H. R. "*Gigantopithecus blacki* von Koenigswald, a Giant Fossil Hominoid from the Pleistocene of Southern China." *Anthropological Papers of the American Museum of Natural History* 43, no. 4 (1952): 295–325.

Woo, Ju-Kang. "The Mandibles and Dentition of *Gigantopithecus.*" *Palaeontologia Sinica* 146, no. 11 (1962): 1–94.

Zhang, Yinyun. "Variability and Evolutionary Trends in Tooth Size of *Gigantopithecus blacki.*" *American Journal of Physical Anthropology* 59, no. 1 (1982): 21–32.

Zhao, L. X., and L. Z. Zhang. "New Fossil Evidence and Diet Analysis of *Gigantopithecus blacki* and Its Distribution and Extinction in South China." *Quaternary International* 286 (2013): 69–74.

14. Breaking Back

Books

Johanson, Donald C., and Maitland A. Edey. *Lucy: Beginnings of Humankind*. Simon & Schuster, 1981.

Articles

Anderson, Robert. "Human Evolution, Low Back Pain, and Dual-Level Control." In *Evolutionary Medicine*, edited by Wenda R. Trevathan, E. O. Smith, and James J. McKenna, 333–49. Oxford University Press, 1999.

Leakey, Mary D. "Tracks and Tools." *Philosophical Transactions of the Royal Society of London. Series B, Biological Sciences* 292, no. 1057 (1981): 95–102.

Lovejoy, C. Owen. "Evolution of Human Walking." *Scientific American* 259, no. 5 (1988): 118–25.

Rosenberg, Karen R., and Wenda R. Trevathan. "Bipedalism and Human Birth: The Obstetrical Dilemma Revisited." *Evolutionary Anthropology* 4, no. 5 (1996): 161–68.

15. In Search of the Most Humanlike Face

Books

Bowman-Kruhm, Mary. *The Leakeys: A Biography*. Prometheus, 2009.

Morell, Virginia. *Ancestral Passions: The Leakey Family and the Quest for Humankind's Beginnings*. Touchstone, 1996.

Articles

Antón, Susan C., Richard Potts, and Leslie C. Aiello. "Evolution of

Early *Homo*: An Integrated Biological Perspective." *Science* 345, no. 6192 (2014): 1236828-1 to -13.

Leakey, Louis S. B. "A New Fossil Skull from Olduvai." *Nature* 184, no. 4685 (1959): 491–93.

Leakey, Louis S. B., Phillip V. Tobias, and J. R. Napier. "A New Species of the Genus *Homo* from Olduvai Gorge." *Nature* 202, no. 4927 (1964): 7–9.

Leakey, Meave G., Fred Spoor, M. Christopher Dean, Craig S. Feibel, Susan C. Antón, Christopher Kiarie, and Louise N. Leakey. "New Fossils from Koobi Fora in Northern Kenya Confirm Taxonomic Diversity in Early *Homo*." *Nature* 488, no. 7410 (2012): 201–4.

Leakey, Richard E. F. "Evidence for an Advanced Plio-Pleistocene Hominid from East Rudolf, Kenya." *Nature* 242, no. 5398 (1973): 447–50.

Wood, Bernard A., and Mark Collard. "The Human Genus." *Science* 284, no. 5411 (1999): 65–71.

16. Our Changing Brains

Books

Lieberman, Daniel E. *The Evolution of the Human Head*. Belknap Press, 2011.

Articles

Aiello, Leslie C., and Robin I. M. Dunbar. "Neocortex Size, Group Size, and the Evolution of Language." *Current Anthropology* 34, no. 2 (1993): 184–93.

Aiello, Leslie C., and Peter E. Wheeler. "The Expensive-Tissue Hypothesis: The Brain and the Digestive System in Human and Primate Evolution." *Current Anthropology* 36, no. 2 (1995): 199–221.

Dunbar, Robin I. M. "Evolution of the Social Brain." *Science* 302, no. 5648 (2003): 1160–61.

Kaplan, Hillard S., and A. J. Robson. "The Emergence of Humans: The Coevolution of Intelligence and Longevity with Intergenerational Transfers." *Proceedings of the National Academy of Sciences of the USA* 99, no. 15 (2002): 10221–26.

Lee, Sang-Hee, and Milford H. Wolpoff. "The Pattern of Pleistocene Human Brain Size Evolution." *Paleobiology* 29, no. 2 (2003): 185–95.

17. You Are a Neanderthal!

Books

Finlayson, Clive. *Humans Who Went Extinct: Why Neanderthals Died Out and We Survived.* Oxford University Press, 2009.

Pääbo, Svante. *Neanderthal Man: In Search of Lost Genomes.* Basic Books, 2014.

Stringer, Christopher, and Clive Gamble. *In Search of the Neanderthals: Solving the Puzzle of Human Origins.* Thames & Hudson, 1994.

Articles

Boule, M. "L'homme fossile de La Chapelle-aux-Saints." *Annales de Paléontologie* 6 (1911–13): 11–172.

D'Errico, Francesco, João Zilhão, Michele Julien, Dominique Baffier, and Jacques Pelegrin. "Neanderthal Acculturation in Western Europe? A Critical Review of the Evidence and Its Interpretation." *Current Anthropology* 39, no. 2 (1998): s1–s44.

Frayer, David W., Ivana Fiore, Carles Lalueza-Fox, Jakov Radovčić, and Luca Bondioli. "Right Handed Neandertals: Vindija and Beyond." *Journal of Anthropological Sciences* 88 (2010): 113–27.

Green, Richard E., Johannes Krause, Adrian W. Briggs, Tomislav Maricic, Udo Stenzel, Martin Kircher, Nick Patterson, et al. "A Draft Sequence of the Neandertal Genome." *Science* 328, no. 5979 (2010): 710–22.

Green, Richard E., Johannes Krause, Susan E. Ptak, Adrian W. Briggs,

Michael T. Ronan, Jan F. Simons, Lei Du, et al. "Analysis of One Million Base Pairs of Neanderthal DNA." *Nature* 444, no. 7117 (2006): 330–36.

Krings, Matthias, Helga Geisert, Ralf W. Schmitz, Heike Krainitzki, and Svante Pääbo. "DNA Sequence of the Mitochondrial Hypervariable Region II from the Neandertal Type Specimen." *Proceedings of the National Academy of Sciences of the USA* 96, no. 10 (1999): 5581–85.

Noonan, James P. "Neanderthal Genomics and the Evolution of Modern Humans." *Genome Research* 20, no. 5 (2010): 547–53.

Stringer, Christopher B. "Documenting the Origin of Modern Humans." In *The Emergence of Modern Humans*, edited by Erik Trinkaus, 67–96. Cambridge University Press, 1989.

Thorne, Alan G., and Milford H. Wolpoff. "The Multiregional Evolution of Humans." *Scientific American* 266, no. 4 (1992): 76–83.

Wolpoff, Milford H. "The Place of the Neandertals in Human Evolution." In *The Emergence of Modern Humans*, edited by Erik Trinkaus, 97–141. Cambridge University Press, 1989.

18. The Molecular Clock Does Not Keep Time

Books

Crow, James F., and Motoo Kimura. *An Introduction to Population Genetics Theory*. Harper and Row, 1970.

Marks, Jonathan. *What It Means to Be 98% Chimpanzee: Apes, People, and Their Genes*. University of California Press, 2002.

Articles

Cann, Rebecca L., Mark Stoneking, and Alan C. Wilson. "Mitochondrial DNA and Human Evolution." *Nature* 325, no. 6099 (1987): 31–436.

Green, Richard E., Johannes Krause, Adrian W. Briggs, Tomislav

Maricic, Udo Stenzel, Martin Kircher, Nick Patterson, et al. "A Draft Sequence of the Neandertal Genome." *Science* 328, no. 5979 (2010): 710–22.

Kimura, Motoo. "Possibility of Extensive Neutral Evolution under Stabilizing Selection with Special Reference to Nonrandom Usage of Synonymous Codons." *Proceedings of the National Academy of Sciences of the USA* 78 (1981): 5773–77.

Krings, Matthias, Helga Geisert, Ralf W. Schmitz, Heike Krainitzki, and Svante Pääbo. "DNA Sequence of the Mitochondrial Hypervariable Region II from the Neandertal Type Specimen." *Proceedings of the National Academy of Sciences of the USA* 96, no. 10 (1999): 5581–85.

Li, Wen-Hsiung, and Lori A. Sadler. "Low Nucleotide Diversity in Man." *Genetics* 129, no. 2 (1991): 513–23.

Wilson, Alan C., and Rebecca L. Cann. "The Recent African Genesis of Humans." *Scientific American* 266, no. 4 (1992): 68–73.

19. Denisovans: The Asian Neanderthals?

Books

Harris, Eugene E. *Ancestors in Our Genome: The New Science of Human Evolution*. Oxford University Press, 2014.

Articles

Hawks, John. "Significance of Neandertal and Denisovan Genomes in Human Evolution." *Annual Review of Anthropology* 42, no. 1 (2013): 433–49.

Huerta-Sánchez, Emilia, Xin Jin, Asan, Zhuoma Bianba, Benjamin M. Peter, Nicolas Vinckenbosch, Yu Liang, et al. "Altitude Adaptation in Tibetans Caused by Introgression of Denisovan-like DNA." *Nature* 512, no. 7513 (2014): 194–97.

Krause, Johannes, Qiaomei Fu, Jeffrey M. Good, Bence Viola, Michael

V. Shunkov, Anatoli P. Derevianko, and Svante Paabo. "The Complete Mitochondrial DNA Genome of an Unknown Hominin from Southern Siberia." *Nature* 464, no. 7290 (2010): 894–97.

Meyer, Matthias, Qiaomei Fu, Ayinuer Aximu-Petri, Isabelle Glocke, Birgit Nickel, Juan-Luis Arsuaga, Ignacio Martinez, et al. "A Mitochondrial Genome Sequence of a Hominin from Sima de los Huesos." *Nature* 505, no. 7483 (2014): 403–6.

Meyer, Matthias, Martin Kircher, Marie-Theres Gansauge, Heng Li, Fernando Racimo, Swapan Mallick, Joshua G. Schraiber, et al. "A High-Coverage Genome Sequence from an Archaic Denisovan Individual." *Science* 338, no. 6104 (2012): 222–26.

20. Hobbits

Books

Morwood, Mike, and Penny Van Oosterzee. *A New Human: The Startling Discovery and Strange Story of the "Hobbits" of Flores, Indonesia.* Smithsonian, 2007.

Articles

Falk, Dean, Charles Hildebolt, Kirk Smith, M. J. Morwood, Thomas Sutikna, Peter J. Brown, Jatmiko, E. Wayhu Saptomo, Barry Brunsden, and Fred Prior. "The Brain of LB1, *Homo floresiensis.*" *Science* 308, no. 5719 (2005): 242–45.

Hayes, Susan, Thomas Sutikna, and Mike Morwood. "Faces of *Homo floresiensis* (LB1)." *Journal of Archaeological Science* 40, no. 12 (2013): 4400–10.

Martin, Robert D., Ann M. MacLarnon, James L. Phillips, and William B. Dobyns. "Flores Hominid: New Species or Microcephalic Dwarf?" *Anatomical Record* 288A, no. 11 (2006): 1123–45.

Morwood, M. J., R. P. Soejono, R. G. Roberts, T. Sutikna, C. S. M. Turney, K. E. Westaway, W. J. Rink, et al. "Archaeology and Age of a

New Hominin from Flores in Eastern Indonesia." Nature 431, no. 7012 (2004): 1087–91.

Van Den Bergh, Gerrit D., Bo Li, Adam Brumm, Rainer Grün, Dida Yurnaldi, Mark W. Moore, Iwan Kurniawan, et al. "Earliest Hominin Occupation of Sulawesi, Indonesia." *Nature* 529, no. 7585 (2016): 208–11.

Wong, Kate. "The Littlest Human." *Scientific American* 292, no. 2, (2005): 56–65.

21. Seven Billion Humans, One Single Race?

Books

Gould, Stephen J. *The Mismeasure of Man*. W. W. Norton, 1981.

Wolpoff, Milford H., and Rachel E. Caspari. *Race and Human Evolution: A Fatal Attraction*. Simon & Schuster, 1997.

Articles

Caspari, Rachel E. "From Types to Populations: A Century of Race, Physical Anthropology, and the American Anthropological Association." *American Anthropologist* 105, no. 1 (2003): 65–76.

Coon, Carleton S. "New Findings on the Origin of Races." *Harper's Magazine* 225, no. 1351 (1962): 65–74.

Day, Michael H., and Christopher B. Stringer. "A Reconsideration of the Omo Kibish Remains and the *erectus-sapiens* Transition." In *L'Homo erectus et la place de l'homme de Tuatavel parmi les hominidés fossiles,* edited by Marie-Antoinette de Lumley, 814–46. Centre National de la Recherche Scientifique, 1982.

Livingstone, Frank B. "On the Non-existence of Human Races." *Current Anthropology* 3, no. 3 (1962): 279.

Stringer, Christopher B., and Peter Andrews. "Genetic and Fossil Evidence for the Origin of Modern Humans." *Science* 239, no. 4845 (1988): 1263–68.

Wolpoff, Milford H. "Describing Anatomically Modern *Homo sapiens*: A Distinction without a Definable Difference." *Anthropos (Brno)* 23 (1986): 41–53.

Wolpoff, Milford H., Xinzhi Wu, and Alan G. Thorne. "Modern *Homo sapiens* Origins: A General Theory of Hominid Evolution Involving the Fossil Evidence from East Asia." In *The Origins of Modern Humans*, edited by Fred H. Smith and Frank Spencer, 411–83. Alan R. Liss, 1984.

22. Are Humans Still Evolving?

Books

Cochran, Gregory M., and Henry Harpending. *The 10,000 Year Explosion: How Civilization Accelerated Human Evolution*. Basic Books, 2009.

White, Leslie A. *The Evolution of Culture: The Development of Civilization to the Fall of Rome*. McGraw-Hill, 1959.

Articles

Frayer, David W. "Metric Dental Change in the European Upper Paleolithic and Mesolithic." *American Journal of Physical Anthropology* 46, no. 1 (1977): 109–20.

Hawks, John, Eric T. Wang, Gregory M. Cochran, Henry C. Harpending, and Robert K. Moyzis. "Recent Acceleration of Human Adaptive Evolution." *Proceedings of the National Academy of Sciences of the USA* 104, no. 52 (2007): 20753–58.

Huerta-Sánchez, Emilia, Xin Jin, Asan, Zhuoma Bianba, Benjamin M. Peter, Nicolas Vinckenbosch, Yu Liang, et al. "Altitude Adaptation in Tibetans Caused by Introgression of Denisovan-like DNA." *Nature* 512, no. 7513 (2014): 194–97.

Yi, Xin, Yu Liang, Emilia Huerta-Sánchez, Xin Jin, Zha Xi Ping Cuo, John E. Pool, Xun Xu, et al. "Sequencing of 50 Human Exomes

Reveals Adaptation to High Altitude." *Science* 329, no. 5987 (2010): 75–78.

Appendix 1. Common Questions and Answers about Evolution

Books

Crow, James F., and Motoo Kimura. *An Introduction to Population Genetics Theory.* Harper and Row, 1970.

Darwin, Charles. *The Descent of Man, and Selection in Relation to Sex.* John Murray, 1871.

Darwin, Charles. *On the Origin of Species.* John Murray, 1859.

Jablonka, Eva, and Marion J. Lamb. *Evolution in Four Dimensions: Genetic, Epigenetic, Behavioral, and Symbolic Variation in the History of Life.* MIT Press, 2006.

Articles

Gould, Stephen Jay, and Niles Eldredge. "Punctuated Equilibria: The Tempo and Mode of Evolution Reconsidered." *Paleobiology* 3, no. 2 (1977): 115–51.

Appendix 2. Overview of Hominin Evolution

Books

Cela-Conde, Camilo J., and Francisco J. Ayala. *Human Evolution: Trails from the Past.* Oxford University Press, 2007.

Johanson, Donald C., and Blake Edgar. *From Lucy to Language.* Simon & Schuster, 1996.

Articles

Green, Richard E., Johannes Krause, Adrian W. Briggs, Tomislav Maricic, Udo Stenzel, Martin Kircher, Nick Patterson, et al. "A

Draft Sequence of the Neandertal Genome." *Science* 328, no. 5979 (2010): 710–22.

Reich, David, Richard E. Green, Martin Kircher, Johannes Krause, Nick Patterson, Eric Y. Durand, Bence Viola, et al. "Genetic History of an Archaic Hominin Group from Denisova Cave in Siberia." *Nature* 468, no. 7327 (2010): 1053–60.

Wolpoff, Milford H., John D. Hawks, David W. Frayer, and Keith Hunley. "Modern Human Ancestry at the Peripheries: A Test of the Replacement Theory." *Science* 291, no. 5502 (2001): 293–97.

Wolpoff, Milford H., Alan G. Thorne, Jan Jelínek, and Yinyun Zhang. "The Case for Sinking *Homo erectus*: 100 Years of *Pithecanthropus* Is Enough!" In *100 Years of Pithecanthropus: The Homo erectus Problem*, edited by J. L. Franzen, 341–61. Frankfurt, 1994.

Wood, Bernard A., and Mark Collard. "The Human Genus." *Science* 284, no. 5411 (1999): 65–71.

Index

Page numbers in *italics* refer to illustrations.

Acheulean hand axes, 72
acquired characteristics,
 inheritance of, 252
adaptation, culture as, 227
adaptive advantage, adaptive
 success, 15–16, 110
 lactase persistence as, 81–82
 mutations and, 252–53
adaptive strategy, 256
adenine (A), 190–91
Afghanistan, 79
Africa:
 environmental change in, 70
 hominin migration out of, 194,
  229, 230, 257, 260–62
 hominin origins in, 47, 121–22
Africans, 90
aging, *see* longevity
agriculture, 105–12, 228, 230, 236
 artificial selection in, 232–33
 and increases in disease and
  malnutrition, 106–9
 population explosions and,
  109–12, 231, 236
 and rise of class systems, 111, 228,
  236
 rise of, 105–7
 skin color and, 91
 supposed benefits of, 105, 236
 warfare and, 110, 236
Aiello, Leslie, 175
Aka people, 210
Altai Mountains, Russia, 199
 hominins in, 201
altruism, 131–39
 in ants and bees, 132–33
 in early hominins, 137–38
 as evolved human trait, 139
 genes and, 133–34
 in monkeys, 132
 in Neanderthals, 135–37, 139
 toward nonrelations, 131–32,
  134–35
 and transmission of cultural
  information, 138–39
Alzheimer's disease, 73
Amazon basin, 26
amino acids, 191, 192, 193
Andaman Islands, 210
Andes, 28
animal husbandry, 228, 236
 artificial selection and, 83, 232–33

animal husbandry (*continued*)
and increase in human disease, 109, 237
lactase persistence and, 79–81
animals, eusocial, 133
Antón, Susan, 116
ants, altruism in, 132–33
apes:
big toes of, 154
evolution of humans from, 253–54
monkeys vs., 253–54
see also primates
APOE4 (apolipoprotein epsilon-4), 72, 74
apolipoprotein, 72
brain disease and, 73–74
Ardipithecus kadabba, 255
Ardipithecus ramidus, 39, 52–53, 255
Arens, William, 24
Armelagos, George, 111
art:
of Neanderthals, 186
Upper Paleolithic blossoming of, 99–100
artificial selection, 83, 232–33
Asia:
Homo erectus in, 201
search for *Australopithecus* fossils in, 200
Asians, 90
light skin of, 230
Atapuerca, Spain, 27
Australia, 208, 209
Australian Aborigines, 209
arrival in Australia of, 219
distinctive look of, 219, 222–23
Australopithecus (genus), 98–99, 122, 137, 199, 255–56, 264
African origins of, 125
brain size of, 172
possible out-of-Africa migration of, 212–13
search for non-African fossils of, 200
see also Paranthropus
Australopithecus aethiopicus, 256
dental morphology of, 256–57
Australopithecus afarensis, 39, 210, 212, 256
Australopithecus anamensis as possible variant of, 50–51
bipedalism of, 50, 153
female pelvis of, 64
radiometric dating of, 49
small brain of, 50, 152
Australopithecus africanus, 44, 49–50, 159, 162, 210, 212, 213, 215, 216, 256
brain size of, 124
discovery of, 124–25
Australopithecus anamensis, 255–56
bipedalism of, 50
as possible *Australopithecus afarensis* variant, 50–51
Australopithecus bahrelghazali, 256
Australopithecus boisei, 256
brain size of, 175
Australopithecus garhi, 256
brain size of, 55
toolmaking by, 55
Australopithecus robustus, 256, 258
Australopithecus sediba, 256

back pain, bipedalism and, 155, 157, 235
bamboo, 146–47
base pairs, 190–91
Beatles, 152
bees, altruism in, 132–33
Berger, Lee, 256*n*
biochemistry, 49
bipedalism, 37, 39, 48, 54, 162, 256, 257
of *Australopithecus afarensis*, 50, 153
of *Australopithecus anamensis*, 50
back and knee pain as consequence of, 155, 157, 235
of birds, 157–58
childbirth and, 60–61
childcare and, 155, 235
of early hominins, 48, 52–53, 54
heart disease and, 155–56, 157, 235
in humans, 157–58
language and, 156
morphological changes associated with, 153
as preceding increased brain size, 152, 154, 156–57
toolmaking and, 156
birds, bipedalism of, 157–58
birth canal, size of, fetal brain size vs., 60–61, 62
birth celebrations, delayed, 94
body size:
Gigantopithecus and, 143
of Gorillas, 142–43
growth period and, 145
in *Homo erectus*, 97
male-male competition and, 143
meat eating and, 72, 144
body size sexual dimorphism, 143, 227–28
in chimpanzees, 36
in *Gigantopithecus*, 143–44
in gorillas, 35
in humans vs. gorillas, 39
bonobos (*Pan paniscus*), recreational sex among, 40
Brace, C. Loring, 85
brain, human:
aging and, 169
of children vs. adults, 170
energy requirements of, 156, 174, 177
lateral asymmetry in, 185
myths about, 169
processing capabilities of, 169
synapses of, 170–71, 176
brain disease, apolipoprotein and, 73–74
brain size:
of *Australopithecus*, 172
of *Australopithecus afarensis*, 50
of *Australopithecus africanus*, 124
of *Australopithecus garhi*, 55
caloric intake and, 71–72
of chimpanzees, 172
cognition and, 156–57
of early hominins, 48, 54, 69, 151, 152, 256
of *Homo*, 257
in *Homo erectus*, 97
of *Homo floresiensis*, 207, 209
in human vs. nonhuman newborns, 60–61
of modern humans, 235–36
of Neanderthals, 172
of social animals, 173, 236

brain size, increase in:
digestive system size and, 175
food addictions as consequence of, 177
hunter-gatherer lifestyle as necessitated by, 175, 236
language and, 151
meat eating and, 71–72, 157, 175
possible reversal of, 176
as preceded by bipedalism, 152, 154, 156–57
skull size and, 175
social brain hypothesis and, 171, 172–77
toolmaking and, 151, 157, 172
breastfeeding, ovulation and, 110
burials:
purposeful, of Neanderthals, 139–40
secondary, 22–23
Burj Khalifa, Dubai, 121

calcium, milk as source of, 82
Calico, California, 13
California, University of, at Riverside, 9
Calment, Jeanne, 101
canine teeth, size of, in male-male competition, 144
Cann, Rebecca, 189–90, 193
cannibalism, 19–31
author's course on, 19–20
Columbus's account of, 23–24
cut marks on bones as evidence of, 21–23, 27–28
in extreme circumstances, 28
false reports of, 24–25
Peking Man fossils and, 118–19
ritualistic, 25–26, 28, 30, 31
carnivores, 69, 70
Caspari, Rachel, 98
Caucasus, 200
Centers for Disease Control and Prevention (CDC), 92
Central Intelligence Agency (CIA), 114
chatting:
as grooming, 174
as primary function of language, 173
cheese, 82
childbirth, in humans:
bipedalism and, 60–61
fetal brain size and, 60–61, 62
hospitals and, 65–66
mother's need for assistance in, 62–63
nonhuman childbirth vs., 59–66
pain of, 59, 157, 236
as social activity, 63–66
as traumatic experience for fetus, 61
childbirth, in nonhumans, 61
mother's need for solitude in, 62–63
childcare:
bipedalism and, 155, 235
longevity and, 96–99, 261
males and, 33–42
modern division of responsibilities in, 42
childhood, malnutrition in, 106–7
chimpanzees, 254
bipedalism of, 157–58
body size sexual dimorphism in, 36
brain size of, 172

divergence of hominins and, 49, 255
as humans' closest living relative, 255
as meat eaters, 68
reproductive strategy among, 36
China, 84, 200
Homo erectus in, 145–46
Japanese invasion of, 114
Chronicle of Higher Education, 20
Ciochon, Russ, 146
civilization, rise of, 228, 236, 237
civil rights movement, 10
class systems, rise of, agriculture and, 111, 228, 236
climate change, 147, 148, 260
Cochran, Gregory, 231
codons, 191, 193
cognition, brain size and, 156–57
Columbus, Christopher, 23–24
competitive advantage, 36
complete replacement model, 194, 197, 223–25, 262, 263
cooperation, 134
environmental change and, 138
as evolved human trait, 139
as necessary for hominin survival, 173, 175–76
couvade syndrome, 42–43
Creutzfeldt-Jakob disease (CJD), 30
Cro-Magnons, 181–82
cultural information, 236
generational transmission of, 100, 138–39, 151, 236
culture:
as adaptation, 227
biological evolution and, 227, 229, 230–31
cytosine (C), 190–91
dairy economies, 84
lactase persistence and, 79–81
Dart, Raymond, 124
Darwin, Charles, 48, 180, 193, 248–49, 250
Dawkins, Richard, 133
Dawson, Charles, 124, 127
deforestation, 146–47, 148
dementia, 73
Denisova cave, Russia, 199, 201–2, 204–5, 261
Denisovans, 199–205, 232*n*, 261, 264
DNA of, 202–3
geographical and temporal range of, 204–5
modern humans' relationship to, 199, 203, 204
Denmark, 79
Dennell, Robin, 127
dental disease, agriculture and, 108
diabetes, 177
digestive system, brain size and, 175
disease, increases in, agriculture and, 106, 108–9, 237
Dmanisi, Georgia, *102*, 126, 127, 137, 212, 260
DNA (deoxyribonucleic acid), 183
base pairs of, 190–91
of Denisovans, 202–3
of Neanderthals, 182–83, 185, 197, 263
noncoding ("junk"), 193, 196, 197–98
primate lineages and, 49
replication of, 191
dolphins, recreational sex among, 40

Dong A Il Bo, 242–43
Donner Party, 28
dragon bones, 141–42
dragons, 141
Dubois, Eugène, 122–24
Dunbar, Robin, 171, 172–73
dwarfism, 209
  insular, 211

Early Pleistocene, 257, 261
East Africa, 49
East Asia, early hominins in, 114
Ebu Gogo, 207
ecology, generational knowledge of, 236
Eldredge, Niles, 253
enamel hypoplasia, 106–7, 147
environmental change, 70, 147, 148, 260
  human impact on, 236–37
  social cooperation and, 138
enzymes, 191
EPAS1, 232
epigenetics, 252
estrus, 34–36, 38, 40
Ethiopia, 55, 152
ethnographic research, 106
Eucharist, 26
eugenics, 218
European early modern humans, 81, 98–99, 181, 221–22, 231
Europeans:
  light skin of, 230
  New World voyages of, 217
  Northern, lactase persistence in, 79, 82, 84
  racist portrayals of indigenous peoples by, 181–82
eusocial animals, 133
evolution:
  as change in gene frequency over generations, 193
  Darwin's theory of, 248–49, 250–51
  definition of, 247
  diversity and, 231
  epigenetics and, 252
  natural selection in, 193, 250–52
  neutral theory of, 192–97, 229, 251–52
  population and, 250
  progress as different from, 252–53
  punctuated equilibrium model of, 253
  reproduction of traits in, 249–50
  role of mutations in, 111–12, 249
  sequential view of, 254
  sexual selection in, 250–51
evolution, human:
  biological and cultural forces in, 227
  complete replacement model of, 223–25, 262, 263
  as continuing process, 83, 227–34
  in descent from apes, 253–54
  ever-changing ideas about, 15
  multiregional model of, 223–25, 262–63
  old model of, 199–200
  racism and, 224–25
  as winding river, 16
evolution, human, accelerated rate of, 229
  biological and cultural forces in, 229, 230–31

genetic exchanges and, 231
medical advances and, 232, 233–34
population explosion and, 231
regional variation and, 232
evolutionary advantage, 95–96
evolutionary biology, 73–74, 110
evolutionary success, 132
expensive tissue hypothesis, 175

face, humanlike, of *Homo rudolfensis*, 161
Falk, Dean, 210, 213, 216
farming, *see* agriculture
fatherhood:
biological changes induced by, 41
as cultural concept, 41–42
evolution of, 33–42
in Lovejoy model, 41
in traditional patriarchies, 42
females:
estrus in, 34–36, 38, 40
pelvis of, 64
postmenopausal life spans of, 96
reproductive capacity of, 34
see also childbirth; childcare
fertility rates, 94, 95
agriculture and, 109
fire, *Homo erectus*'s use of, 116–17
Flores Island, 207–11
evidence of hominins on, 208
stone tools found on, 210
fluorine dating, 128
folic acid, 89
Fore people:
kuru among, 29–30, 31
ritual cannibalism practiced by, 25–26, 28, 30, 31
fossils:
difficulty of determining species of, 221, 222, 261–62
morphology of, 60, 137, 141, 163, 182, 203, 209–10, 211, 220, 256
Foster, Jodie, 19
FOXP2, 185
Frayer, David, 185, 231
fur, as mammalian characteristic, 86

Gajdusek, Daniel, 30, 31
Galilei, Galileo, 248
genes:
altruism and, 133
exchanges of, and accelerated rate of evolution, 231–31
longevity and, 95
multiregional evolution model and, 224
mutations of, *see* mutations
skin color and, 90–91, 229–30
genetic adaptation, 73
genetic determinism, 134
genetic diversity:
agriculture and, 111–12
population explosions and, 112
genetics, 49
basics of, 190–91
growing impact on paleoanthropology of, 264
see also epigenetics; population genetics
genomes:
comparison of, 220
of Neanderthals vs. modern humans, 197
sequencing of, 228–29

geocentrism, heliocentrism vs., 248
Germany, Neanderthal heritage celebrated in, 186
giants, in mythology, 148–49
gibbons, 254
Gigantopithecus, 141–49, 207
  author's research on, 144
  body size of, 143
  body size sexual dimorphism in, 143–44
  climate change and, 147
  dental disease in, 147
  habitat of, 146–47
  human competition as leading to extinction of, 146–47
  predators and, 145
  southern Chinese range of, 145
  teeth and mandibles as only known fossils of, 142
Gigantopithecus blacki, 142
giraffes, 252
Gods Must Be Crazy, The (film), 106
Gona, Ethiopia, 64
gorillas:
  bipedalism of, 157–58
  body size of, 142–43
  dental morphology of, 256
  hominin divergence from, 49
  reproductive strategy among, 34–35
"Got Milk?" (advertising campaign), 77, 83–84
Gould, Stephen Jay, 253
grandmother hypothesis, 96–99, 261
  author's testing of, 98–99
gratitude relay, 235
grooming, chatting as, 174
guanine (G), 190–91
Gutenberg University, 81
Gwa Hak Dong A, 14, 241–43

Hadar, Ethiopia, 49
hair:
  as advantage for early hunters, 87
  evolution of, 87–88
  as human characteristic, 86–87
  reduction in extent in, 88
Hamilton, William, 133
Hamilton's law, 133–34
handedness:
  language and, 185–86
  of Neanderthals, 185–86
Harpending, Henry, 231
Harvard Kalahari Project, 106
Hawkes, Kirsten, 96
Hawks, John, 231
health, longevity and, 93–94
heart disease, 177
  bipedalism and, 155–56, 157, 235
heliocentrism, geocentrism vs., 248
herbivores, 68–69
heritability, of acquired characteristics, 252
heroism, 131–32
Holloway, Ralph, 210, 213
hominins:
  in Altai region, 201
  bipedalism in, *see* bipedalism
  divergence from chimpanzees and gorillas of, 49, 255
  genetic exchanges by, 232
  grandmother hypothesis and, 98–99

out-of-Africa migration of, 229, 230, 257, 260–62
toolmaking by, 201, 235
hominins, early, 13, 28, 37, 39, 70, 162, 255–56
African origins of, 121–22
altruism in, 137–38
brain size of, 48, 54, 69, 151, 152, 256
in East Asia, 114
environmental change and, 70
evolution of, 47–55
hair of, *see* hair
hunting by, 87, 173
increasing brain size in, 60
male-male competition in, 39
as meat eaters, 67
plant-based diet of, 68–69, 70
reproductive strategy in, 39
skin color and, 229
social cooperation as necessary for survival of, 173
sweating by, 88
teeth size of, 48, 54
toolmaking by, 48, 54
Homo (genus), 69, 162
brain size of, 257
as hunters, 67, 73
meat eating by, 67–68, 69, 257
as scavengers of meat, 70–71
Homo antecessor, 261
Homo cepranensis, 261
Homo erectus, 69, 162, 199, 212, 260, 261–62, 264
African- vs. Asian-origin hypothesis for, 121–26, 127–28
in Asia, 201
body size of, 97
brain size of, 97, 175
in China, 145–46
female pelvis of, 64
fire as used by, 116–17
grandmother hypothesis and, 97–98
hunting by, 145–46, 257
Java Man as, 123, *129*
as meat eater, 72–73
out-of-Africa migration of, 212
Peking Man as, 116, 125
possible bamboo tools and shelters of, 146–47
toolmaking by, 257
Homo ergaster, see Homo erectus
Homo floresiensis, 207–8, 216, 261
brain size of, 207, 209
possible toolmaking in, 210, 211
small stature of, 207, 210
trapezoid bone of, 211
Homo georgicus, *102*, 260, 261
Homo habilis, 69, 72, 163–64, 167–68, 257, 264
brain size of, 175
toolmaking by, 54, 56, 71
variation in specimens of, 165
Homo heidelbergensis, 261
Homo naledi, 256*n*
Homo neanderthalensis, see Neanderthals
Homo rhodesiensis, 261
Homo rudolfensis, 69, 161, 165, 166, 167, 257
toolmaking by, 71
Homo sapiens, see humans, modern
Homo sapiens idaltu, 262
Hopkins, Anthony, 19
horses, *Homo erectus* as hunters of, 145–46

human genome, sequencing of, 228–29
humans, ancestral, *see* hominins
humans, modern (*Homo sapiens*), *18*, 165
  brain size of, *see* brain size
  chimpanzees as closest living relatives of, 255
  debate on origins of, 223–24
  definition of, 222
  Denisovans' relationship to, 199, 203, 204
  early European, 81, 98–99, 181, 221–22, 231
  environmental impact of, 236–37
  evolution of, *see* evolution, human
  hair of, *see* hair
  as meat lovers, 67–68, 69
  monogamy among, *see* monogamy
  Neanderthal gene sequences in, 183, 185, 203
  Neanderthals compared to, 221–22
  Neanderthals' interbreeding with, 189–90, 223, 241, 263
  Neanderthals' relationship to, 180–81, 199
  as omnivores, 20–21
  as predators, 237
  as social animals, 63, 65, 236
  see also Neanderthals
hunter-gatherers, 236
  early hominins as, 87, 173
  early *Homo* as, 67, 73
  Homo erectus as, 145–46, 257
  interbirth interval among, 109–10, 261
  large brain size of, 175
  revised view of, 106
  social cooperation as necessary for survival of, 173, 175–76
hybrid species, hybridization, 219*n*
hypervitaminosis, 69–70

Ice Age, 89, 91, 116–17, 136, 137
  environmental change in, 138
India, 84
Indiana, Pennsylvania, 10–11
Indiana University of Pennsylvania, 10
indigenous peoples, Europeans' racist portrayals of, 181–82
Indonesia, 114, 122–23, 207–11, 261
infant mortality, 94
information, sharing of:
  big brain as necessity for, 176, 236
  by hunter-gatherers, 175–76
  intergenerational, 100, 138–39, 151, 236
insular dwarfism, 211
interbirth interval, 109–10, 261
Ivanov, Ilya Ivanovich, 220

Jackman, Hugh, 83
Java, 123, 148, 208
Java, Indonesia, 261
Java Man, 114, 122–24, 260, 261
  as *Homo erectus*, 123, *129*
Jebel Irhoud, Morocco, *18*
Jesus Christ, 26
Johanson, Donald, 49, 152
Jordan, 79
junk (noncoding) DNA, 193, 196, 197–98
Jurassic Park (film), 183

Kalahari people, 106, 261
Kang, Jaegu, 131–32
Kansas, 13
Kara-Bom, Russia, 201
Kennedy, John F., 186
Kentucky, 11
Kenya, 161, 162
Kenyanthropus platyops, 256
Kimura, Motoo, 191
kinship, fictive, as uniquely human trait, 134–35
KNM-EM 62000, 161
KNM-ER 1470, 164
  classification of, 165–66, 167
KNM-ER 1808 (*Homo erectus* fossil), 69, 73
KNM-ER 62000, 166–67
Koenigswald, Gustav Heinrich Ralph von, 141
Koobi Fora, Kenya, 69, 161, 163, 166
Korea, 10, 43
  agriculture in, 105
  average life span in, 93
  delayed birth celebrations in, 94
  heroic depictions of ancestors in, 186
  racism in, 186
  Sewol ship tragedy in, 131–32
Krapina, Croatia, Neanderthal burials at, 21–23, *31*
!Kung people, 106, 261
kuru, 29–30, 31
kwashiorkor, 107

labor, sexual division of, 39
La Chapelle-aux-Saints, France, 135–36, 181, *184*
lactase, 77–78
  as needed by infants, 78
lactase deficiency, 77–79
  as global norm in adults, 78–79, 84
  viewed as disease in U.S., 78
lactase persistence:
  as abnormal condition, 79
  as adaptive advantage, 81–82
  diary economies and, 79–81
  gene mutations for, 80, 83
  in Middle East, 82
  in Northern Europeans, 79, 82, 84
lactation, ovulation and, 110
lactational amenorrhea, 110
lactose, 77–78
Laetoli, Tanzania, 49
  hominin footprints at, 154, 162
Lamarck, Jean-Baptiste, 252
language:
  bipedalism and, 156
  brain size and, 151
  chatting as primary function of, 173–74
  FOXP2 gene and, 185
  Neanderthals and, 185
Late Pleistocene, 204
Late Pliocene, 256, 257
lateral asymmetry, 185
LCT (lactase gene), 80
Leakey, Louis, 13, 54, 162–63, 164, 167
Leakey, Mary, 49, 54, 162–63, 164
Leakey, Meave, 167
Leakey, Richard, 164, 166, 167
Leakey family, 161–62
Lee, Sang-Hee, *238*
  cannibalism course of, 19–20
  cross-country trip of, 9–14
  evolving teaching methods of, 14–15

Lee, Sang-Hee (*continued*)
Gigantopithecus research of, 144
grandmother hypothesis tested by, 98–99
gratitude list of, 235–37
Korean magazine and newspaper columns of, 14–15, 241–43
life span:
absolute, 101
increase in, 93, 94, 95, 100–101
see also longevity
limb bones, length of, agriculture and, 107
Li, Wen-Hsiung, 193
Livingstone, Frank, 85–86
longevity, 93–101, 236
childcare and (grandmother hypothesis), 96–99, 261
health issues and, 93–94
heritability of, 95
in human evolution, 96–99
increase in, 94–96, 100–101, 112
of postmenopausal women, 96
socioeconomic implications of, 94, 236
and transmission of cultural information, 100, 138–39, 236
Lord of the Rings, The (Tolkien), 207
Lovejoy, Owen, 39, 154
Lovejoy model, 39–40, 261
fatherhood in, 41–42
"Low Nucleotide Diversity in Man" (Li and Sadler), 193–94
Lucy (*Australopithecus afarensis*), 49, 64, 152, 210
Lucy (film), 169
"Lucy in the Sky with Diamonds" (song), 152
McDonald's, 177
mad cow disease, 30
male-male competition, 34–35, 39
body size and, 143
canine tooth size and, 144
males:
as fathers, 33–42
reproductive success of, 33–34
sympathetic pregnancy (couvade syndrome) in, 42–43
malnutrition, increases in, agriculture and, 106–7
mammals, fur of, 86
Man-Eating Myth, The (Arens), 24
Marvel Comics, 191
masticatory muscle, skull size and, 175
Max Planck Institute, 182, 196
meat eating, 67–74
body size and, 72
brain size and, 71–72, 157, 175
by *Homo*, 67–68, 69, 257
Medicare, 94
medicine, advances in, and accelerated rate of evolution, 232, 233–34
Meganthropus, 148–49
Melanesia, Melanesians, 203, 204
melanin, 85, 88, 89, 90, 229, 257
menopause, 96
Mesolithic, 231
Mezmaiskaya cave, Russia, 200–201, 202
Michigan, 11
Michigan, University of, 14
microcephaly, 209
Middle East, 79
lactase persistence in, 82

Middle Pleistocene, 261
climate fluctuation in, 147
migrations, human:
out of Africa, 194, 212–13, 229, 230, 257, 260–62
skin color and, 89, 91
milk:
as calcium and protein source, 82
effect of animal husbandry and artificial selection on, 83
increase in global consumption of, 84
intolerance to, *see* lactase deficiency; lactase persistence
as vitamin D source, 82
missing link, 253
mitochondrial DNA, 189, 194–95, 202
faulty assumptions about, 196
low diversity of, 193–94
molecular clocks, 189–90
as based on faulty assumptions, 195
Cann's use of, 189–90
Mongolia, 199
monkeys:
altruism in, 132
apes vs., 253–54
monogamy, in humans, 38
male biological changes induced by, 41
morphology:
of fossil bones, 60, 137, 141, 163, 182, 203, 209–10, 211, 220, 256
of modern humans, 222, 231, 232
mortality rates, 94, 95, 110
agriculture and, 109
mortality risks, cycle of, 94–95
Morwood, Michael, 207, 208, 213
"Mother's Heart" (song), 59
Moula-Guercy, France, 27
Mousterian tools, 200
mules, infertility of, 219
multiregional evolution model, 11, 223–25, 262–63
mutations, 191–92
adaptive advantage and, 252–53
beneficial vs. harmful, 192–93
evolutionary role of, 111–12, 249
frequency of, 193
neutral theory of, 192–97, 229, 251–52
of noncoding DNA, 196
reproductive fitness and, 192

National Geographic Society, 115
Native Americans:
evidence of possible cannibalism by, 27–28
secondary burials of, 22–23
natural selection, 72, 74, 111, 193, 196, 228–29, 230, 249, 250–52
Nature, 161, 166, 189
nature, respect for, 236
Neanderthals, *31*, 60, 98–99, 179–85, *184*, 186, 194, 261
altruism in, 135–37, 139
bone cut marks as evidence for cannibalism by, 21–23, 27
brain size of, 172
cave art of, 186
classification of, 200
discovery of, 180
DNA from, 182–83, 185, 197, 263
full genome of, 263
gene sequences from, in modern humans, 183, 185, 203

Neanderthals (*continued*)
genome of, 197
geographical and temporal range of, 179*n*, 200–201, 205
interbreeding between modern humans and, 189–90, 223, 241, 263
language and, 185
modern humans compared to, 221–22
modern humans' relationship to, 180–81, 199
negative image of, 134–35, 179–81, 186
newborn skull of, 64
purposeful burial of, 139–40
teeth of, 185–86
toolmaking by, 186, 200
Neolithic Age, 80–81, 83
neutral theory of mutations, 192–97, 229, 251–52
Newtown, Connecticut, 131
New World, European discovery of, 217
New York Times, 161
nuclear family, 39, 40
nutrients, nutrition:
regulation of, skin color and, 89–91
see also malnutrition

Oldowan tools, 55, 56, 71, 210
Olduvai Gorge, Tanzania, 162–63
Olorgesailie, Kenya, 75
ominivores, 69
"On the Non-existence of Human Races" (Livingstone), 86
On the Origin of Species (Darwin), 180, 217, 249, 251
orangutans, 147
"Origin of Man, The" (Lovejoy), 39
Orrorin tugenensis, 51–52, 53, 121, 255
out-of-Africa migrations, 194, 229, 230, 257, 260–62
skin color and, 89, 91
overeating, and brain size increase, 177
ovulation, lactation and, 110

Pääbo, Svante, 182, 183, 189, 241
paegil (delayed birth celebrations), 94
paleoanthropology, 11, 13
growing impact of genetics on, 264
Paleolithic Period, Upper, *see* Upper Paleolithic
Papua New Guinea, 25–26, 203
Paranthropus, 70
Paranthropus boisei, see Australopithecus boisei
Paranthropus robustus, see Australopithecus robustus
pathogens:
evolution of, agriculture and, 108–9, 237
zoonotic (species-jumping), 109, 237
patriarchies, traditional, role of fathers in, 42
peacocks, 250
Peking Man, 125, 261
as *Homo erectus*, 116, 125
Peking Man fossils, 125
cannibalism and, 118–19
casts of, 115–16, 125

disappearance of, 113–15
lack of facial bones in, 118–19
yakuza and, 113–14, 115, 117–18
Piltdown Man, 48, 124, 127–28
Pithecanthropus erectus, see Homo erectus
pleiotropy hypothesis, 73–74
Pleistocene, 27, 28, 204, 231, 257, 261, 262
environmental change in, 70, 147
Pliocene, 211, 255, 256, 257
Polynesians, 90
Ponce de León, Marcia, 64
population:
aging of, 95
evolution and, 250
population explosions, 236, 260
and accelerated rate of evolution, 231
agriculture and, 109–12, 231
genetic diversity and, 112
population genetics, 192–93, 251–52
complete replacement model of, 194
neutral theory of, 192–94
predators, prey body size and, 144
primates:
childbirth in, 61
child rearing by, 33–42
grooming by, 174
lineages of, 49
male-male competition in, 34–35, 39
plant-based diet of, 68
see also apes; *specific species*
prions, 30, 31
Proconsul, 48–49
progress, evolution as different from, 252–53
protein, 191, 192, 236
insufficient intake of, 107
milk as source of, 82
PTC, 221
punctuated equilibrium model, 253

race:
as biological concept, lack of evidence for, 218
origin of concept of, 217–18
skin color and, 85–86, 218, 221
as social construct, 86, 221
species and, 218–19, 221
racial inequality, 10
racism:
in portrayals of indigenous peoples and Neanderthals, 181–82
and theory of human evolution, 224–25
radiocarbon dating, 128
Ramapithecus, 48–49
regional traits, 232, 261–63
Reich, David, 91, 230
relaxin, 61
reproduction, of morphological variations, 249–50
reproductive fitness, mutations and, 192
reproductive strategy, 34–36, 39
reproductive success, 33, 36, 96, 111
child rearing and, 37
as goal of all life-forms, 59–60
Richards, Keith, 19, 20
Rihanna, 83

Rocky Mountains, 13
Rosenburg, Karen, 62
Russell, Mary, 22–23, 27
Russia, 199

Sadler, Lori, 193
Sahelanthropus tchadensis, 51–52, 53, 121, 255
Sandy Hook Elementary School shooting, 131
San people, 106, 261
Sarich, Vince, 49
Science, 39, 52, 64
secondary burials, 22–23
selective advantage, 229, 234
Selfish Gene, The (Dawkins), 133
Sewol ship tragedy, 131–32
sex, viewed as social concept, 228
sexual activity, recreational, 38, 40
sexual dimorphism, body size, *see* body size sexual dimorphism
sexual selection, 250–51, 253
Shanidar 1, 137, 139
Shanidar 4, 139–40
Shanidar cave, Iraq, 137, 139–40
Sierra Nevada, 28
Silence of the Lambs, The (film), 19
Sinanthropus pekinensis, *see* Peking Man
skin cancer, 92
skin color, 257
  agriculture and, 91
  of early hominins, 229
  as genetically determined, 85, 90–91, 229–30
  geographic distribution of, 90
  human migrations and, 89, 91
  nutrient regulation and, 89–91
  race and, 85–86, 218, 221
  ultraviolet radiation and, 86, 88–89
  vitamin D hypothesis of, 229–30
skull size:
  increase in brain size and, 175
  masticatory muscle and, 175
social animals, brain size of, 173, 236
social brain hypothesis, 171, 172–73
social life, as hominin survival tactic, 173, 175–76
Social Security, 94
sociobiology, 133
Sociobiology (Wilson), 133
Solomon Islands, 203
Southeast Asia, Southeast Asians, 186, 208
speciation, isolation as condition for, 218–19, 220–21
species:
  biological definition of, 219, 262, 263–64
  of fossils, difficulty of determining, 221, 222, 261–62
  genomic comparison of, 220
  race and, 218–19, 221
  variation within, 165, 232, 261–63
Spurlock, Morgan, 177
Star Trek: Voyager (TV show), 10
Steinbeck, John, 9–10
Sterkfontein, South Africa, *44*, *159*
Stewart, Jimmy, 10
stone tools, 54, 55, *56*, 75, 151, 172
Stringer, Christopher, 222, 223
stroke, 73
subspecies, 219–20, 221

Sudan, 79, 80, 82
sunscreens, overuse of, vitamin D deficiency and, 92
Super Size Me (film), 177
Survive! (film), 28
Swartkrans, South Africa, 258
Sweden, 79, 80
synapses, 176
  in human brain, 170–71

tanning salons, 92
Taung, South Africa, 215
Taung Child (*Australopithecus africanus*), 124, 210, 213, *215*, 216
teachers, Korean vs. U.S. attitudes toward, 14
teeth:
  size of, 48, 54
  wisdom, 233–34
10,000 Year Explosion, The (Cochran and Harpending), 231
testosterone, 40, 41
Thorne, Alan G., 224
thymine (T), 190–91
Tibetans, 204
Tolkien, J. R. R., 207
toolmaking, tool use, 48, *56*, *75*, 163, 235
  in Altai region, 201
  bipedalism and, 156
  brain size and, 151, 172
  by *Australopithecus garhi*, 54–55
  by *Homo erectus*, 257
  by *Homo floresiensis*, 210, 211
  by *Homo habilis*, 54, 71
  by *Homo rudolfensis*, 71
  by Neanderthals, 186, 200
Toumaï, Chad, 51
traits:
  as inherently neither advantageous nor disadvantageous, 233
  regional, 232, 261–63
  reproduction of, 249–50
trapezoid bone, 211
Travels with Charley (Steinbeck), 9–10
Trevathan, Wenda, 62
Tugen Hills, Kenya, 51

ultraviolet radiation, 92, 257
  dangers of, 88
  skin color and, 86, 88–89, 229
  in vitamin D synthesis, 82, 89–91, 229, 230
  vitamin D synthesis and, 82, 89–91, 229
United States:
  dairy economy in, 84
  lactase deficiency in, 78, 79, 83–84
University College London, 81
Upper Paleolithic, 22, 201, 222, 231
  blossoming of art and symbols in, 99–100
  emergence of human longevity in, 98–99
Uruguayan rugby team, 28
Ust-Karakol, Russia, 201

variation:
  reproduction of, 249–50
  within species, 165, 232, 261–63
Vietnam War, 11
Vindija cave, Croatia, 202
vitamin A, 69–70

vitamin D, 229
in milk, 82
sunscreen overuse and, 92
UV radiation in synthesis of, 82, 89–91, 229, 230
voyages of discovery, 217

Wallace Line, 208, 209
warfare, agriculture and, 110, 236
Weidenreich, Franz, 115–16, 125, 148
Wheeler, Peter, 175
White, Leslie, 227
White, Tim, 154
Wilson, Allan, 49
Wilson, Edward O., 132
wisdom teeth, 233–34
Wolpoff, Milford, 11, 176, 222, 223, 224
Wu, Xinzhi, 224

X-Men, 191

yakuza, missing Peking Man fossils and, 113–14, 115, 117–18
Yanomami, 26
yogurt, 82
Yoon, Shin-Young, 14, 237, 241, 244

Zhang, Yinyun, 144
Zhoukoudian cave, China, 113, 114, 116, 118, 125, 145, 261
Zika virus, 209
Zinjanthropus (genus), *see also Australopithecus*
Zinjanthropus boisei, 162, 163, 164
Zollikofer, Christoph, 64
zoonoses (species-jumping pathogens), 109, 237